Guide to Tendrillate Climbers of Costa Rican Mountains

Alexander Krings is curator of the North Carolina State University Herbarium.

Richard R. Braham is professor of forestry, College of Natural Resources, North Carolina State University-Raleigh. He is coauthor of *North American Trees, 5th Edition*, a popular textbook of tree identification.

Blackwell Publishing Professional
2121 State Avenue, Ames, Iowa 50014, USA

Orders: 1-800-862-6657
Office: 1-515-292-0140
Fax: 1-515-292-3348
Web site: www.blackwellprofessional.com

Blackwell Publishing Ltd
9600 Garsington Road, Oxford OX4 2DQ, UK
Tel.: +44 (0)1865 776868

Blackwell Publishing Asia
550 Swanston Street, Carlton, Victoria 3053, Australia
Tel.: +61 (0)3 8359 1011

First edition, 2005

Library of Congress Cataloging-in-Publication Data

Krings, Alexander.
Guide to tendrillate climbers of Costa Rican mountains / Alexander Krings, Richard R. Braham.—1st ed.
p. cm.
Includes bibliographical references (p.) and index.
ISBN 0-8138-0758-1 (alk. paper)
1. Climbing plants—Costa Rica—Identification. 2. Forest plants—Costa Rica—Identification.
3. Mountain plants—Costa Rica—Identification. I. Braham, Richard R. II. Title.

QK217.K75 2005
582.1′8′097286—dc22

2004015352

The last digit is the print number: 9 8 7 6 5 4 3 2 1

CONTENTS

of specimens or from previously published articles, floras, or monographs. With few exceptions each entry also includes information on the timing of flowering and fruiting, gathered from our field observations and dates on herbarium specimens collected by others. The phenology data are incomplete, but they provide a useful beginning. Information on the overall geographic distribution and typical elevational range in Costa Rica are also provided. The distributional and elevational data are also preliminary. Additional notes are provided for selected species.

Each species is illustrated by a figure drawn by Alexander Krings. The collector, collection number, and name of the herbarium that contains the specimen serving as the basis for each illustration is indicated in the caption of the illustration. The standard abbreviations defined above are used for herbaria. The amount of information provided in each entry varies, reflecting our sometimes incomplete current state of knowledge. We hope that this guide will direct the study of others to improve our knowledge of these important plants.

Terminology for gross plant morphology largely follows Harris and Harris (1994). With some exceptions, abbreviations are from the following sources. Journals follow Lawrence et al. (1968) and Bridson and Smith (1991). Books follow Stafleu and Cowan (1976), authors of scientific names follow Brummitt and Powell (1992), and herbaria follow Holmgren et al. (1990).

This guide could not have been written without help from Pedro Acevedo Rodríguez, Shaun Bennett, Laura Bolaños, William Burger, Robert Carlson, the staff of Casa Ridgway, Susan Cohen, the staff of the Costa Rican and North American offices of the Organization of Tropical Studies, Frederick Cubbage, Christian Feuillet, Luis Diego Gómez Pignataro, Michael Grayum, William Haber, Bruce Haines, Barry Hammel, James Hardin, David Hearn, Gail Hewson, Humberto Jiménez-Saa, Richard Lancia, the staff at Las Cruces Biological Station, Silvia Lobo, John MacDougal, the Monteverde Conservation League, Aisha Mora, Alejandra Murillo, Ronald Murillo, Alejandra Okie, Christopher Peterson, Pedro Rosafa, the staff at La Casona, Armando Ruiz, Theodore Shear, Marsha Sitnik, Brenda Sneed, James Solomon, David Spratte, Donald Stone, Larry Tombaugh, Daniel Turk, Kevyn Wightman, Robert Wilbur, Wendy Worley, and Nelson Zamora. Jeremy Mendel scanned the illustrations.

We would also like to expressly thank the curators and staffs at CR, DUKE, F, INB, MO, NCSC, US, and USJ for their assistance. Loans of specimens from F and MO were very helpful.

This work was supported by grants provided by the Departments of Forestry and Botany at North Carolina State University, the Organization for Tropical Studies, the Field Museum of Natural History, and the Smithsonian Institution. This generous support is gratefully acknowledged.

Guide to Tendrillate Climbers of Costa Rican Mountains

stitute less than 1% of European species, but they make up about 19% of neotropical species (Gentry 1991). Although the Neotropics may contain the greatest diversity of plants on earth (Raven 1976; Gentry 1988), African forests harbor the greatest number of liana species followed in order of decreasing diversity by the Neotropics and Asia (Emmons and Gentry 1983; Gentry 1991).

Four somewhat discontinuous mountain ranges in Costa Rica strongly influence the vegetation, species richness, and structure and distribution of climber communities. In general, lowland plant communities contain more species and more widespread species than mountain communities (Gentry 1991, 1995). Conversely, mountain communities contain more species with restricted distributions (Gentry 1986). This higher level of endemism can possibly be explained by the isolating effects of discontinuous mountain ranges.

The main mountain ranges are the Cordillera de Guanacaste in the northwest, Cordillera de Tilaran and Cordillera Central in the middle of the country, and Cordillera de Talamanca in the southeast. These mountains separate the Caribbean and Pacific lowlands, and present a discontinuous barrier mostly ranging from 500 to 3,000 m in height and stretching almost the entire length of Costa Rica. Only in the northwest near the Nicaraguan border does this barrier drop below 400 m (Burger 1985). About 43% of Costa Rica lies above 700 m and is classified as mountainous.

As a consequence of the prevailing winds, the western slope of all four ranges is distinctly seasonal, and may support seasonally deciduous forests. The eastern slope receives abundant rainfall throughout most of the year, and supports tropical evergreen forests. Depending upon elevation, the mountains themselves support either premontane, lower montane, or montane vegetation (Holdridge et al. 1971; Hartshorn 1983). Among the most striking of mountain communities are Cloud Forests, which are almost continually covered in fog, and as a result contain trees covered with epiphytes. Depending upon the humidity at the base of the mountain, Cloud Forests are found at elevations ranging from 1,000 to 2,500+ m.

In montane Costa Rica, most plant families containing climbers utilize only one of four basic climbing mechanisms. Only five families (Bignoniaceae, Polygonaceae, Rhamnaceae, Rubiaceae, and Solanaceae) utilize two, and only one (Fabaceae) uses three. The most common climbing mechanism is twining, successively wrapping the main stem completely around the host plant. The following 27 or 28 families contain twiners: Acanthaceae, Amaranthaceae, Apocynaceae (*s.l.*), Aristolochiaceae, Asteraceae, Basellaceae, Boraginaceae, Celastraceae, Combretaceae, Connaraceae, Convolvulaceae, Ericaceae, Euphorbiaceae, Fabaceae, Gentianaceae, Grossulariaceae, Hippocrateaceae, Malpighiaceae, Phytolaccaceae, Polygalaceae, Polygonaceae, Ranunculaceae, Rubiaceae, Solanaceae, Tropaeolaceae, Valerianaceae, Verbenaceae, and possibly Campanulaceae.

Climbing by tendrils is the second most common mechanism in Costa Rica with the 11 families listed in the preface table P.1. Tendrillate climbers use specialized tissues that wrap around leaves or small lateral branches of the host plant. Several tissues have been modified into tendrils, including terminal leaflets (Bignoniaceae), terminal buds (Cucurbitaceae and Vitaceae), axillary buds (*Bauhinia*, *Strychnos*, *Passiflora*), terminal portions of inflorescences (*Antigonon*), and stipules (*Smilax*) (figs. 1.1 and 1.2). Tendrillate climbers often become established when the host tree is small, and remain in the crown by matching the height growth of the host tree. However, some tendrillate climbers, such as *Macfadyena unguis-cati*, develop hooklike tendrils that allow them to ascend large-diameter trees. Other taxa (e.g., *Cissus* and *Vitis*) form many-branched tendrils, and are capable of ascending large trees by lodging the tendril tips in bark crevices. Tendril position and structure are very useful identification characters. In conjunction with leaf arrangement, structure, and venation, they readily distinguish the 11 tendrillate families of montane Costa Rica.

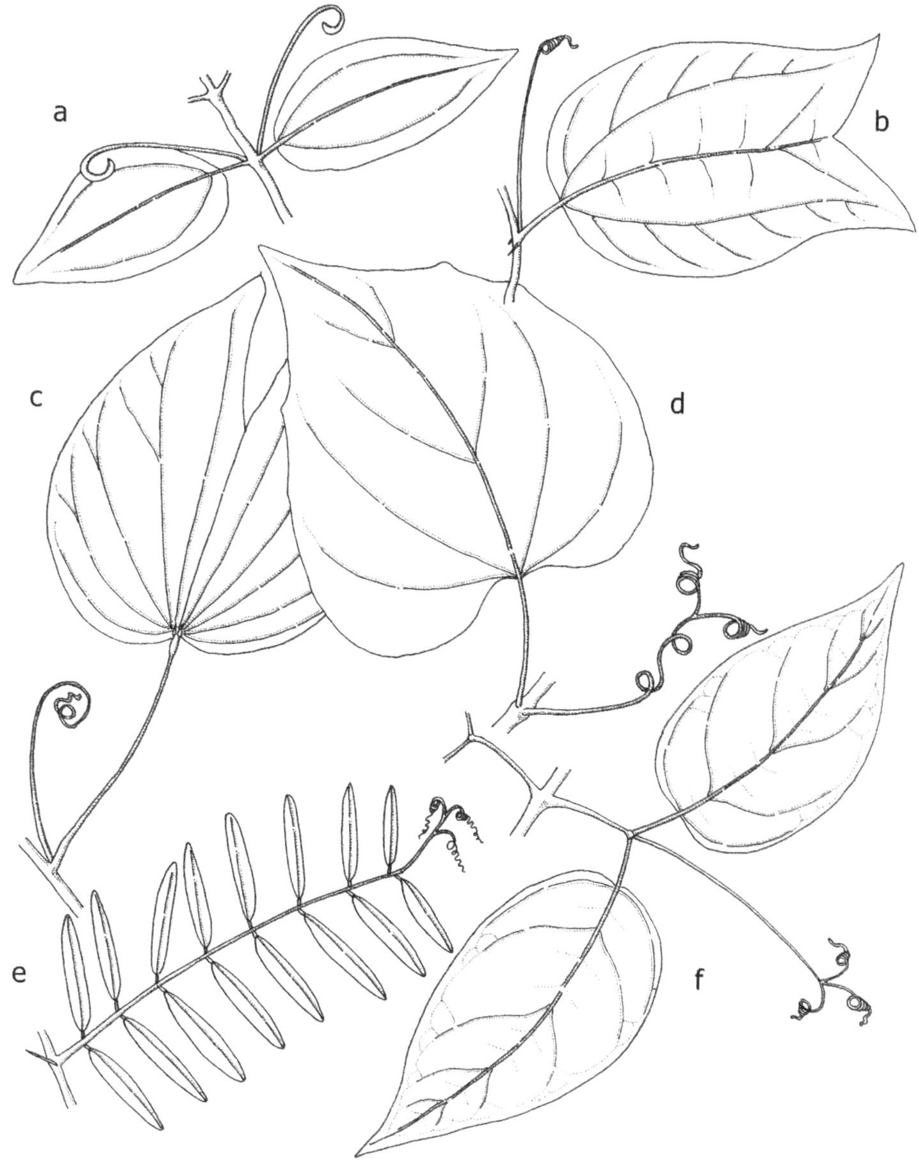

Fig. 1.1. Tendril position in families Bignoniaceae to Passifloraceae: **(a)** Loganiaceae (*Strychnos*: axillary), **(b)** Passifloraceae (*Passiflora*: axillary), **(c)** Fabaceae (*Bauhinia*: axillary), **(d)** Cucurbitaceae (at right angles to petiole), **(e)** Fabaceae (*Vicia*: terminating leaf), **(f)** Bignoniaceae (replacing terminal leaflet).

Tendrils are the most specialized of all climbing adaptations (Darwin 1867). About 30% of the climbers in the mountains of Costa Rica utilize tendrils, and about 20% of Costa Rican dicot families with climbers contain tendrillate species. Tendrils have arisen independently in many families, but are much more common among dicotyledons than monocotyledons. Among monocots, only *Smilax* (Smilacaceae) is tendrillate. No tendrillate gymnosperms are known.

2
FAMILIES

KEY TO FAMILIES

1. Leaves opposite
   2. Leaves simple; tendril often hooklike, axillary to a leaf or bract... Loganiaceae (*Strychnos*), p. 79.
   2. Leaves compound; tendril often in place of terminal leaflet... Bignoniaceae, p. 11.
1. Leaves alternate
   3. Tendrils in pairs, stipular, arising at base of petiole... Smilacaceae (*Smilax*), p. 151.
   3. Tendrils not in pairs, not arising from base of petiole
      4. Tendril opposite leaf at node... Vitaceae, p. 161.
      4. Tendril not opposite leaf at node
         5. Tendril arising at 90-degree angle to petiole... Cucurbitaceae, p. 37.
         5. Tendril not arising at 90-degree angle to petiole
            6. Leaves compound
               7. Leaves 2-foliolate... Fabaceae (*Bauhinia*), p. 75.
               7. Leaves not 2-foliolate
                  8. Tendril axillary... Sapindaceae, p. 129.
                  8. Tendril terminating the rachis
                     9. Leaves 4–8 jugate or bipinnately compound... Fabaceae (*Entada* and *Vicia*), p. 75.
                     9. Leaves 2–3 jugate... Polemoniaceae (*Cobaea*), p. 119.
            6. Leaves simple
               10. Tendrils borne at the end of the inflorescence... Polygonaceae (*Antigonon*), p. 123.
               10. Tendrils not borne at the end of inflorescence
                   11. Mature stems flattened, sigmoid to undulate; leaves cleft; venation arcuate... Fabaceae (*Bauhinia*), p. 75.
                   11. Mature stems terete, mostly straight; leaves not cleft; venation not arcuate
                       12. Tendril axillary, but not at apex of short branch, coiled in loose spiral... Passifloraceae (*Passiflora*), p. 87.
                       12. Tendril axillary to leaf at apex of short branch, coiled tightly like a watch spring... Rhamnaceae (*Gouania*), p. 125.

3
BIGNONIACEAE

A largely tropical family of trees, shrubs, and lianas, the Bignoniaceae contains over 120 genera and 650 species (Richardson 1978). One of the most important liana families of neotropical lowland forests (Gentry 1974a), the Bignoniaceae is not endemic to the Neotropics. A few members occur also in Africa and Asia (Gentry 1982). In the Neotropics, the Bignoniaceae is generally widespread and found in such diverse habitats as dry forests, lowland wet forests, and montane forests (Gentry 1985). Absolute density of the Bignoniaceae decreases from dry to wet forests, while species richness increases (Gentry 1991).

The family is characterized by large, showy flowers and wind-dispersed diaspores (Gentry 1974a, 1993; Croat 1978). It is divided into three major tribes (Bignonieae, Crescentieae, and Tecomeae) and several minor tribes. Bignonieae contains roughly 400 species of lianas, and is generally considered the most advanced (Gentry 1974b, 1991). The tribes Crescentieae and Tecomeae are largely arborescent.

Tribal divisions are based primarily on fruit morphology. Crescentieae has indehiscent fruits, whereas Bignonieae and Tecomeae have dehiscent capsules. In Bignonieae, capsules dehisce parallel to the septum, but in Tecomeae the capsules dehisce perpendicular to the septum (Gentry 1974b). The flowers are characterized by two pairs of stamens inserted near the base of the corolla tube with the divergent anther thecae of each pair held together inside the roof of the tube (Gentry 1993).

The family as a whole is distinct in its usually opposite, compound leaves. Lianas in this family are unique vegetatively because the terminal leaflet has been modified into a simple or branched tendril. No other family has opposite, compound leaves with tendrils. The only other family of tendrillate climbers with opposite leaves is the Loganiaceae, which is readily distinguished by simple, more or less three-veined leaves and relatively short hooklike tendrils.

In our area the Bignoniaceae is represented by 24 tendrillate species in 17 genera. Twenty-five additional tendrillate species are found in the lower elevations of Costa Rica, but do not occur in the mountains. The great majority of species in Costa Rican mountains are woody. Only *Tourrettia lappacea* is herbaceous, and it is easily recognized by the serrate, biternate leaves. The most species-rich genera are *Arrabidaea* and *Mansoa*, each with 3 species. *Amphilophium*, *Anemopaegma*, and *Lundia* have 2 species each. All other genera have 1 species each.

Synonymy follows Gentry (1973), except for *Arrabidaea verrucosa,* which follows Gentry (1977) and *Pithecoctenium crucigerum,* which follows Gentry (1975).

KEY TO BIGNONIACEAE

1. Leaves 2- or 3-ternate; tendrils trifid to many-branched, never simple
   2. Plant woody; leaflet margins entire; tendrils trifid... *Pleonotoma variabilis*, p. 33.
   2. Plant herbaceous; leaflet margins serrate; tendrils finely many-branched... *Tourrettia lappacea*, p. 34.
1. Leaves usually 2–3- or 4–5-foliolate; tendrils simple to many-branched
   3. Leaves palmately 4- or 5-foliolate... *Anemopaegma orbiculatum*, p. 18.
   3. Leaves 2- to 3-foliolate

4. Gland field present near adaxial apex of petiole; tendrils bifid or trifid (sometimes minutely so)... *Paragonia pyramidata*, p. 32.
4. Gland field absent from adaxial apex of petiole (except sometimes present in simple-tendrillate *Arrabidaea verrucosa*); tendrils simple or branched
   5. Stems regularly or slightly irregularly hexagonal
      6. Plants mostly cultivated, possibly naturalized; leaflets mostly rounded or truncate at base; corolla orange or red... *Pyrostegia venusta*, p. 34.
      6. Plants wild; leaflets mostly cordate or subcordate at base; corolla white, yellowish white, or white turning deep purple
         7. Leaf trichomes dendroid or forked; tendrils trifid; calyx double... *Amphilophium*, p. 16.
         7. Leaf trichomes simple; tendrils trifid to many-branched; calyx simple... *Pithecoctenium crucigerum*, p. 33.
   5. Stems terete or tetragonal
      8. Vegetative parts with onion or garlic odor
         9. Tendrils simple; stems subtetragonal to tetragonal; interpetiolar gland fields absent... *Cydista aequinoctialis*, p. 22.
         9. Tendrils trifid; stems subterete to subtetragonal; interpetiolar gland fields present (except in *Mansoa verrucifera*)... *Mansoa*, p. 28.
      8. Vegetative parts lacking onion or garlic odor
         10. Leaves tannish pubescent to velutinous below... *Tynanthus macranthus*, p. 36.
         10. Leaves glabrous to variously pubescent below, but the pubescence not tannish
             11. Stems subtetragonal to tetragonal, with 8 phloem arms in cross-section; interpetiolar gland fields absent; tendrils simple... *Cydista aequinoctialis*, p. 22.
             11. Stems terete, subterete or slightly angular, with 4–many phloem arms in cross-section; interpetiolar gland fields present or absent; tendrils simple or branching
                 12. Interpetiolar gland fields present; stem with 4 phloem arms in cross-section
                     13. Anthers glabrous; disc present; capsule verrucose to tuberculate; leaf bases rounded to truncate... *Arrabidaea verrucosa*, p. 20.
                     13. Anthers pubescent; disc absent or inconspicuous; capsule pubescent to velutinous with median nerve and margins raised; leaf bases cordate to truncate... *Lundia*, p. 25.
                 12. Interpetiolar gland fields absent; stem irregularly many phloem-armed, or with 4–8 phloem arms (unknown in *Tynanthus macranthus*)
                     14. Tendrils simple
                         15. Leaflets with more or less dendroid pubescence below in axils of major lateral veins; interpetiolar gland fields absent; corolla yellow... *Callichlamys latifolia*, p. 22.
                         15. Leaflets lacking dendroid pubescence below in axils of major lateral veins; interpetiolar gland fields present or absent; corolla whitish, magenta, or lavender

16. Stems with 4 phloem arms in cross-section; corolla magenta, lavender, or whitish with magenta ridges... *Arrabidaea*, p. 19.
16. Stems with 8 phloem arms in cross-section; corolla usually white to lavender or magenta... *Cydista aequinoctialis*, p. 22.

14. Tendrils bifid or trifid
    17. Tendrils not uncinately trifid
        18. Stem with interpetiolar ridge connecting opposing leaves; corolla deep wine-red... *Martinella obovata*, p. 30.
        18. Stem lacking interpetiolar ridge; corolla yellow or white
            19. Corolla yellow... *Anemopaegma*, p. 18.
            19. Corolla white... *Distictella magnoliifolia*, p. 23.
    17. Tendrils uncinately trifid
        20. Leaves pubescent to velutinous below; corolla white... *Tynanthus macranthus*, p. 36.
        20. Leaves glabrous below or puberulent along principal veins; corolla yellow
            21. Plant dimorphic, juveniles often rooting at the nodes; stem lacking interpetiolar ridge connecting opposing leaves; tendrils frequently remaining herbaceous with age; capsules to 100 cm long and 1.8 cm wide... *Macfadyena unguis-cati*, p. 26.
            21. Plant monomorphic, not rooting at the nodes; stem with conspicuous interpetiolar ridge connecting opposing leaves; tendrils becoming woody with age; capsules to 15 cm long and 5 cm wide... *Melloa quadrivalvis*, p. 30.

For additional help, tables 3.1–3.3 group our species of Bignoniaceae according to various diagnostic characters.

Table 3.1. Bignoniaceae Taxa Grouped by Vegetative Characters

Character	Taxa
Stems hexagonal	*Amphilophium* (trichomes usually forked or dendroid; tendrils trifid) *Pithecoctenium crucigerum* (trichomes simple; tendrils trifid to many-branching) *Pyrostegia venusta* (stem often irregularly hexagonal or merely angular)
Vegetative parts with onion or garlic odor	*Mansoa* (tendrils trifid in our area) *Cydista* (tendrils simple)
Tendrils simple	*Anemopaegma orbiculatum* (corolla yellow; corolla lobes cream) *Arrabidaea* (corolla magenta to lavender) *Callichlamys latifolia* (corolla yellow with red lines in throat) *Cydista aequinoctialis* (corolla white or lavender, sometimes maroon lines at base of lobes) *Lundia corymbifera* (corolla white with throat ridges yellow) *Lundia puberula* (corolla pinkish to lavender or magenta) *Paragonia pyramidata* (corolla pale lavender to magenta)
Tendrils bifid, trifid, or many-branched	*Amphilophium* (corolla purple) *Anemopaegma chrysanthum* (corolla bright yellow) *Anemopaegma orbiculatum* (corolla yellow; corolla lobes cream) *Distictella magnoliifolia* (corolla white) *Macfadyena unguis-cati* (corolla yellow) *Mansoa hymenaea* (corolla whitish to lavender or magenta) *Mansoa standleyi* (corolla yellow; corolla lobes sometimes red) *Mansoa verrucifera* (corolla purple) *Martinella obovata* (corolla red wine-colored) *Melloa quadrivalvis* (corolla yellow) *Paragonia pyramidata* (corolla pale lavender to magenta) *Pithecoctenium crucigerum* (corolla white, sometimes with yellowish throat) *Pyrostegia venusta* (corolla orange to red) *Tourrettia lappacea* (corolla purplish-greenish) *Tynanthus macranthus* (corolla white or cream)

Table 3.2. Bignoniaceae Taxa Grouped by Corolla Color

Corolla color	Taxa
Red or orange	*Pyrostegia venusta* [*Mansoa standleyi* (corolla yellow; corolla lobes red)]
Yellow	*Anemopaegma* *Callichlamys latifolia* *Macfadyena unguis-cati* *Mansoa standleyi* (sometimes lobes pinkish to rose) *Melloa quadrivalvis* *Pithecoctenium crucigerum* (corolla white with yellow throat)
White, pinkish, purple, lavender to magenta	*Amphilophium* *Arrabidaea* *Cydista* *Distictella* *Lundia* *Mansoa* (except *M. standleyi,* which has a yellow tube with red lobes) *Paragonia pyramidata* *Pithecoctenium crucigerum* (corolla white with yellow throat) *Tourrettia lappacea* (corolla purplish-greenish) *Tynanthus macranthus*

Table 3.3. Bignoniaceae Taxa Grouped by the Number of Arms in the Primary Phloem As Seen in Cross-Section

Number of primary phloem arms	Taxa
4 arms	*Anemopaegma chrysanthum* *Arrabidaea* *Callichlamys latifolia* *Lundia puberula* *Martinella obovata* *Paragonia pyramidata* *Pithecoctenium crucigerum*
4–5 arms	*Amphilophium paniculatum*
4–8 arms	*Mansoa hymenaea*
8 arms	*Anemopaegma orbiculatum* *Cydista aequinoctialis* *Mansoa standleyi* *Mansoa verrucifera*
Irregularly many-arms	*Macfadyena unguis-cati* *Melloa quadrivalvis*
Unknown	*Amphilophium pannosum* *Distictella magnoliifolia* *Pyrostegia venusta* *Tynanthus macranthus*

1. GENUS *AMPHILOPHIUM*

Amphilophium is identified by sharply hexagonal stems and trifid tendrils. It can be distinguished from *Pithecoctenium*, the only other genus with sharply hexagonal stems, by dendroid trichomes on vegetative parts (those of *Pithecoctenium* are simple). *Amphilophium* is represented in Costa Rica by two species that can be differentiated vegetatively by pubescence and persistence of the pseudostipules. Neither species generally contains interpetiolar gland fields.

KEY TO *AMPHILOPHIUM*

1. Stems and leaves lepidote to variously short dendroid pubescent; pseudostipules caducous; corolla 2.4–3.5 cm long; longer filaments to 1.8 cm; shorter filaments to 1.5 cm; capsule smooth to rugose... *A. paniculatum*, p. 16.
1. Stems and leaves at least partly villous with trichomes to 2 mm; pseudostipules persistent; corolla 2.2–2.6 cm long; longer filaments to 1.6 cm; shorter filaments to 1.3 cm; capsule rugose-tuberculate... *A. pannosum*, p. 16.

1. ***Amphilophium paniculatum*** (L.) Kunth in H.B.K., Nov. Gen. sp. pl. 3: 116. 1819. Figure 3.1A.

Synonymy. *Bignonia paniculata* L., Sp. pl. 623. 1753; *A. molle* Schltdl. & Cham., Linnaea 5: 120. 1830; *Endoloma purpurea* Raf., Sylva tellur. 80. 1838; *A. paniculatum* var. *molle* (Schltdl. & Cham.) Standl., Publ. Field Mus. Nat. Hist., Bot. Ser. 18: 1114. 1938.

Description. Vine or liana; stems with 4–5 phloem arms in cross-section; branchlets lepidote to dendroid pubescent, distinctly hexagonal with angles ribbed; interpetiolar gland fields lacking. Leaves bi- or trifoliolate; terminal leaflet sometimes replaced by a tendril; leaflets ovate to suborbicular, 2.5–16.5 cm long, 2–11 cm wide, lepidote above and below, variously dendroid or simply pubescent above and below, palmately veined at base; tip acuminate to obtuse; base truncate to obliquely cordate; margins entire. Tendrils trifid. Inflorescence a panicle, terminal on short lateral shoots. Flowers with double calyx, 0.7–1.2 cm long, 0.8–1.6 cm wide; inner calyx 2–3-lobed; outer calyx more or less 5-lobed, densely lepidote sometimes dendroid pubescent; corolla yellowish white turning deep purple, 2.4–3.5 cm long, 0.7–1.6 cm wide, bilabiate, tubular; corolla lobes 5, viscid, generally glabrous internally, glabrous to short pubescent externally. Fruit a capsule, to 10 cm long and 4 cm wide, ellipsoid, somewhat flattened; surface lenticellate, lepidote. Seeds with membranous wings, nearly 2 cm long and 3 cm wide.

Phenology. Flowers from June through February. Fruits from September through April.

Distribution. Mexico to Argentina. In Costa Rica found from sea level to 2,000 m.

2. ***Amphilophium pannosum*** (DC.) Bureau & K. Schum., in Mart., Fl. Bras. 8: 209. 1896. Figure 3.1B.

Synonymy. *Bignonia pannosa* DC., Prodr. 9: 148. 1845; *B. pubescens* Pav. ex DC., Prodr. 9: 148. 1845, pro syn., non L.; *Amphilophium oxylophium* Donn. Sm., Bot. Gaz. 31: 119. 1901.

Description. Vine or liana; stem cross-section of phloem unknown; branchlets lepidote, villous with simple to dendroid hairs, distinctly hexagonal with angles ribbed; interpetiolar gland fields lacking. Leaves bi- or trifoliolate; terminal leaflet sometimes replaced by a tendril; leaflets ovate, to 13 cm long and 9 cm wide, lepidote and variously dendroid or simply

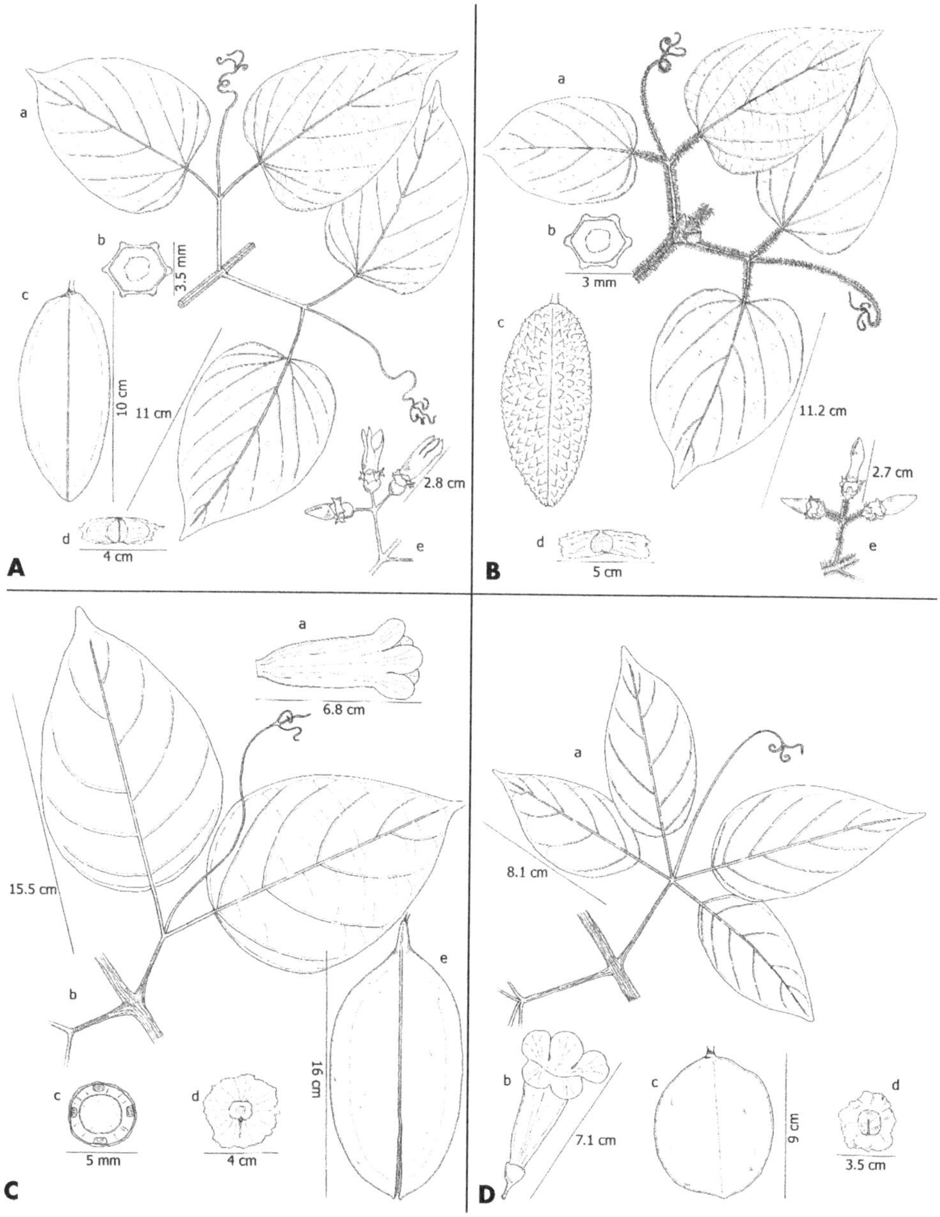

Fig. 3.1. A. *Amphilophium paniculatum:* **(a)** leaves and tendril, **(b)** stem cross-section, **(c)** fruit, **(d)** seed, **(e)** inflorescence (Leaves, stem cross-section, and inflorescence based on Chacón 963, DUKE; Fruit and seed based on Wendt 3835, MO). **B.** *Amphilophium pannosum:* **(a)** leaves, pseudostipules, and tendrils; **(b)** stem cross-section; **(c)** fruit; **(d)** seed; **(e)** inflorescence (Leaves, pseudostipules, stem cross-section, and inflorescence based upon Webster 21942, DUKE; Fruit based on Krings, s.n., Dept. of Forestry, NCSU; Seed based on Quesada 382, MO). **C.** *Anemopaegma chrysanthum:* **(a)** corolla, **(b)** leaf and tendril, **(c)** stem cross-section, **(d)** seed, **(e)** fruit (Leaf, tendril, stem cross-section, and corolla based on Gentry and Dwyer 9374, DUKE; Fruit based on Herréra 2139, MO; Seed based on Beaman 5144, MO). **D.** *Anemopaegma orbiculatum:* **(a)** leaf and tendril, **(b)** flower, **(c)** fruit, **(d)** seed (Leaf and flower based on Jacobs and Grayum 2961, DUKE; Fruit and seed based on Ortíz 1824, MO).

villous above and below, but generally more strongly dendroid villous below, palmately veined; tip acuminate; base subcordate to cordate; margins entire. Tendrils trifid. Inflorescence a racemose panicle, terminal on short lateral shoot. Flowers with double calyx; inner calyx 2–3-lobed; outer calyx more or less 5-lobed, to 1.4 cm long and 1.9 cm wide, densely lepidote, densely dendroid villous below lobes; corolla yellowish white turning deep purple, 2.2–2.6 cm long, 0.7–9 mm wide, bilabiate, tubular; corolla lobes 5, viscid. Fruit a capsule, to 8 cm long and 4 cm wide, ellipsoid, compressed, densely rugose-tuberculate; trichomes simple and dendroid. Seeds with membranous wings, nearly 1.4 cm long and 5 cm wide.

Phenology. Flowers and fruits throughout the year.

Distribution. Costa Rica to Argentina. In Costa Rica found from 200 to 1,700 m.

2. GENUS *ANEMOPAEGMA*

Anemopaegma is represented in Costa Rica by four species, two of which reach our elevations. Our species can be separated primarily by stem and leaf characters. The seeds of both are surrounded by a circular wing, a character distinguishing them from every other genus in our area. Seeds from *Pithecoctenium crucigerum* are the most similar, but frequently are more broadly elliptical than circular. When the seeds are nearly circular, the densely echinate capsule of *Pithecoctenium* separates the two genera.

KEY TO *ANEMOPAEGMA*

1. Stem with 4 phloem arms in cross-section; leaves 2–3-foliolate; fruit oblong-elliptic... *A. chrysanthum*, p. 18.
1. Stem with 8 phloem arms in cross-section; leaves palmately 4–5-foliolate; fruit suborbicular... *A. orbiculatum*, p. 18.

1. ***Anemopaegma chrysanthum*** Dugand, Caldasia 4: 307. 1947. Figure 3.1C.

Description. Liana; stems with 4 phloem arms in cross-section; branchlets subtetragonal, glabrous; interpetiolar gland fields lacking. Leaves bi- to trifoliolate, to 20 cm long and 13 cm wide; terminal leaflet sometimes replaced by a tendril; leaflets ovate to elliptic-ovate, minutely lepidote below; tip acuminate or acute; base cuneate to rounded; margins entire. Tendrils trifid. Inflorescence racemose, axillary, ebracteate. Calyx cupular, 5–9 mm long and wide, more or less lepidote, scattered plate-shaped glands near apex. Corolla tubular-campanulate, 5–7 cm long, 1.2–2.3 cm wide, yellow; tube glabrous externally; lobes to 1.3 cm long, glandular lepidote internally and externally. Fruit a capsule, to 20 cm long and 8 cm wide, ellipsoid, compressed, glabrous. Seeds surrounded by a circular membranous wing, to 5.6 cm long and wide.

Phenology. Flowers from February to May. Fruits from May to November. This species flowers in the dry season (Gentry 1973).

Distribution. Mexico to Colombia and Ecuador. In Costa Rica found from sea level to 1,000 m.

2. ***Anemopaegma orbiculatum*** (Jacq.) DC., Prodr. 9: 190. 1845. Figure 3.1D.

Synonymy. *Bignonia orbiculata* Jacq., Select. stirp. amer. hist. 184, tab. 180, fig. 79. 1763; *Jacaranda orbiculata* (Jacq.) Spreng., Syst. veg. 2: 834. 1825; *Platolaria flavescens* Raf.,

Sylva tellur. 78. 1838; *Pithecoctenium panamense* Benth., Bot. voy. Sulphur 129. 1844; *Peltospermum orbiculatum* (Jacq.) DC., Prodr. 9: 191. 1845, pro. syn.

Description. Liana; stems with eight phloem arms in cross-section; branchlets subtetragonal or subterete, puberulous; interpetiolar gland fields lacking; interpetiolar ridge sometimes present. Leaves palmately 4–5-foliolate; terminal leaflet sometimes replaced by a tendril; leaflets ovate, 4–20 cm long, 2–10 cm wide, minutely lepidote and papillose above and below; tip acuminate; base subcordate to rounded; margins entire. Tendrils simple or trifid. Inflorescence racemose, axillary, puberulous. Calyx cupular, 5–8 mm long and wide, puberulous, sometimes with glands. Corolla tubular-campanulate, 2.5–4 cm long; tube pubescent (especially near point of stamen insertion) to scattered lepidote internally, yellow externally; lobes cream colored, to 1 cm long, glandular-lepidote externally. Fruit a capsule, 7–10 cm wide, orbicular, compressed. Seeds surrounded by a circular, membranous wing, 2.7–4.5 cm wide.

Phenology. Flowers from May to September. Fruits from April to November.

Distribution. Nicaragua to Colombia. In Costa Rica found from 150 to 1,100 m.

3. GENUS *ARRABIDAEA*

The largest genus of bignoniaceous lianas in the world (about 75 species), *Arrabidaea* also contains the most species in Costa Rica. Of nine species that occur in our area, three reach the higher elevations. The genus is recognized by the combination of simple tendrils; flattened, linear capsules; magenta to whitish corollas; and sometimes the presence of interpetiolar gland fields. However, in our area only *Arrabidaea verrucosa* has gland fields.

Arrabidaea most resembles *Lundia*, but can be distinguished by the rounded to truncate leaf bases (mostly cordate but occasionally truncate in *Lundia*). Our other two species of *Arrabidaea* are distinguished from *Lundia* by the lack of interpetiolar gland fields.

KEY TO *ARRABIDAEA*

1. Interpetiolar gland fields present; calyx bilabiate, usually longer than 1.5 cm; capsule verrucose or tuberculate... *A. verrucosa*, p. 20.
1. Interpetiolar gland fields generally absent; calyx usually not bilabiate, generally shorter than 0.9 cm; capsule neither verrucose nor tuberculate
    2. Corolla tube pubescent externally; flower buds uniformly colored; calyx cupular, enclosing corolla base; leaves bi- or trifoliolate... *A. florida*, p. 19.
    2. Corolla tube glabrous externally; flower buds dark with whitish tip; calyx reflexed; leaves simple or bifoliolate... *A. patellifera*, p. 20.

1. ***Arrabidaea florida*** DC., Prodr. 9: 184. 1845. Figure 3.2A.

Synonymy. *Vasconcellia florida* Mart. ex DC., Prodr. 9: 184. 1845, pro syn.; *Arrabidaea panamensis* Sprague, Bull. Herb. Boissier, Sér. 2. 6: 371. 1906.

Description. Liana; stems with four phloem arms in cross-section; branchlets terete, sometimes scabrous but soon glabrate; interpetiolar gland fields usually absent. Leaves bi- or trifoliolate; terminal leaflet sometimes replaced by a tendril; leaflets ovate, to 14 cm long and 7 cm wide, generally puberulous on the veins above and below, sometimes scattered lepidote below, drying to shades of olive or gray; tip acuminate to acute; base cuneate to truncate; margins entire. Tendrils simple. Inflorescence a dense corymb, axillary or terminal. Calyx cupular, 3–4 mm long and 2–3 mm wide, eglandular, puberulous. Corolla magenta to lavender, to 1.7 cm long; tube white internally, to 1.2 cm long; lobes pubescent, to 6 mm long. Fruit a capsule, 10–20 cm long, about 1 cm wide, linear, flattened, sometimes minutely glandular-pitted; midrib and margins often slightly raised. Seeds winged, to 1 cm long and 3.5 cm wide.

Phenology. Flowers from May to November. Fruits from July to December.

Distribution. Belize to Bolivia. In Costa Rica found from 100 to 1,100 m.

2. ***Arrabidaea patellifera*** (Schltdl.) Sandwith, Kew Bull. 22: 413. 1968. Figure 3.2B.

Synonymy. *Bignonia patellifera* Schltdl., Linnaea 8: 516. 1833; *Petastoma patelliferum* (Schltdl.) Miers, Proc. Roy. Hort. Soc. London 3: 195. 1863; *Cuspidaria mollis* Kraenzl., Notizbl. Bot. Gar. Berlin-Dahlem 6: 376. 1915; *Petastoma breviflorum* Standl., J. Arnold Arbor. 11: 128. 1930; *P. pentstemonoides* (Kraenzl.) Sandwith, Kew Bull. 43: 7. 1959; *Arrabidaea broadwayi* (Sprague & Riley) Sandwith, Kew Bull. 22: 414. 1968.

Description. Liana; stems with four phloem arms in cross-section; branchlets terete, glabrous to lepidote, sometimes scabrous; interpetiolar gland fields absent, nodes sometimes with an interpetiolar ridge. Leaves bifoliolate, sometimes simple, sometimes with a tendril; leaflets more or less ovate, to 15 cm long and 11 cm wide, puberulous mostly on veins above and below; tip acuminate to obtuse; base cuneate to rounded; margins entire. Tendrils simple. Inflorescence a panicle, axillary or terminal. Calyx patelliform, 1–4 mm long, 4–7 mm wide, eglandular, scattered lepidote and pubescent. Corolla magenta to lavender, to 4.3 cm long, campanulate above narrowed base; throat frequently white with magenta ridges; tube glabrous internally (except for trichomes near level of stamen insertion), glabrate or with scattered lepidote scales or trichomes externally, to 2.6 cm long; lobes pubescent, to 1.3 cm long. Fruit a capsule, 10–40 cm long, 1–1.5 cm wide, linear, smooth, sometimes with scattered trichomes or lepidote scales, flattened; midrib usually slightly raised; margins slightly raised. Seeds winged, to 1 cm long, nearly 4 cm wide.

Phenology. Flowers from May to October. Fruits from January to November.

Distribution. Mexico to Brazil. In Costa Rica found from 100 to 1,400 m.

3. ***Arrabidaea verrucosa*** (Standl.) A.H. Gentry, Selbyana 2: 43. 1977. Figure 3.2C.

Synonymy. *Paragonia schumanniana* Loes., Bot. Jahrb. Syst. 23: 130. 1897; *Adenocalymma verrucosum* Standl., Publ. Field Mus. Nat. Hist., Bot. Ser. 4: 323. 1929; *Arrabidaea belizensis* Standl., Publ. Field Mus. Nat. Hist., Bot. Ser. 8: 48. 1930; *Martinella verrucosa* (Standl.) Standl., Contr. Arnold Arbor. 5: 138. 1933; *Scobinaria verrucosa* (Standl.) Seibert, Publ. Carnegie Inst. Wash. 522: 408. 1940.

Description. Liana; stems with four phloem arms in cross-section; branchlets terete or subtetragonal, glabrous to puberulous or lepidote; interpetiolar gland fields present. Leaves bifoliolate (sometimes trifoliolate); terminal leaflet sometimes replaced by a tendril; leaflets more or less ovate, to 17 cm long, to 11 cm wide, puberulous on the midvein above, pubescent below, often with simple, long trichomes in axils of veins; tip acuminate to obtuse, base cuneate to rounded; margins entire. Tendrils simple. Inflorescence a panicle, axillary. Calyx tubular-campanulate, 1.7–2.5 mm long, to 1.2 mm wide, strongly bilabiate, lepidote (sometimes only at base). Corolla tubular-campanulate, 4.5–8 cm long, magenta to lavender; throat white, to 5.7 cm long, glabrous internally (except pubescent near point of stamen insertion); lobes pubescent, to 1.8 cm long, sometimes ciliate or fringed. Fruit a capsule, 10–47 cm long, 1.2–2.5 cm wide, linear, flattened, verrucose-tuberculate (tubercules sometimes short). Seeds winged, to 1.8 cm long and 5.2 cm wide.

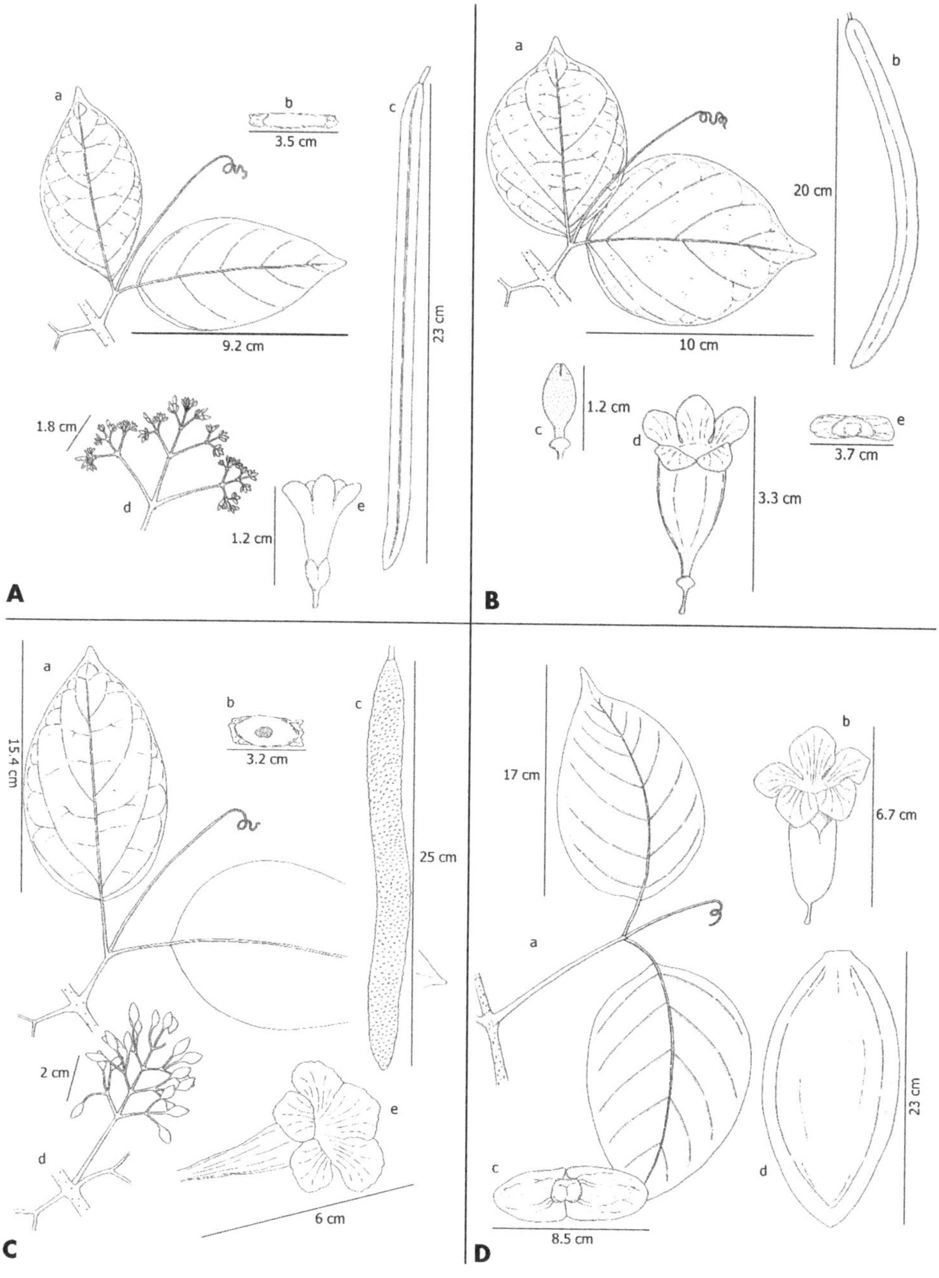

Fig. 3.2. A. *Arrabidaea florida:* **(a)** leaf and tendril, **(b)** seed, **(c)** fruit, **(d)** inflorescence, **(e)** flower (Leaf, tendril, inflorescence, and flower based on de Nevers et al. 5587, DUKE; Fruit and seed based on Liesner and Dwyer 1558, MO). **B.** *Arrabidaea patellifera:* **(a)** leaf and tendril, **(b)** fruit, **(c)** flower bud, **(d)** flower, **(e)** seed (Leaf, tendril, flower bud, and flower based on Burger and Burger 7878, DUKE; Fruit and seed based on Moreno 25440, MO). **C.** *Arrabidaea verrucosa:* **(a)** leaf and tendril, **(b)** seed, **(c)** fruit, **(d)** inflorescence, **(e)** corolla (Leaf, tendril, inflorescence, and corolla based on Hammel 8622, DUKE; Fruit and seed based on Gentry 8163, MO). **D.** *Callichlamys latifolia:* **(a)** leaf and tendril, **(b)** flower, **(c)** seed, **(d)** fruit (Leaf and flower based on Grayum 1298, DUKE; Fruit and seed based on Robleto 1941, MO).

Phenology. Flowers from June through November. Fruits from November through July.
Distribution. Belize to Venezuela and Bolivia. In Costa Rica found from near sea level to 1,200 m.

4. GENUS *CALLICHLAMYS*

Only one species of *Callichlamys* is known from Central America, *C. latifolia*. Useful identification characteristics include dendroid tufts of hair found on leaf undersurfaces in the vein axils; large leaves, corollas, and capsules; and smooth gray bark.

1. *Callichlamys latifolia* (Rich.) K. Schum. in Engler & Prantl, Nat. Pflanzenfam. 4: 223. 1894. Figure 3.2D.

Synonymy. *Bignonia latifolia* Rich., Actes Soc. Hist. Nat. Paris 1: 110. 1792; *Tabebuia latifolia* (Rich.) DC., Rev. Bign. (Bibl. Univ. Geneve) 15. 1838; *Delostoma latifolia* (Rich.) Splitg., Tijdschr. Natuurl. Gesch. Physiol. 9: 11. 1842; *Callichlamys riparia* Miq., Linnaea 18: 254, 609. 1844; *Bignonia crucifera* Bertero ex DC., Prodr. 9: 213. 1845; *Callichlamys rufinervis* (DC.) Miers, Proc. Roy. Hort. Soc. London 3: 201. 1863; *Tabebuia speciosa* Standl., Publ. Field Columbian Mus., Bot. Ser. 8: 49. 1930; *Callichlamys garnieri* Standl. & L.O. Williams, Ceiba 3: 130. 1952.
Description. Liana; stems with four phloem arms in cross-section; branchlets subterete, pubescent to puberulous to glabrate; interpetiolar gland fields absent. Leaves bi- or trifoliolate; terminal leaflet sometimes replaced by a tendril; leaflets more or less ovate or elliptic, to 30 cm long (often less), to 20 cm wide (often less), papillose or lepidote above, generally pubescent below in secondary vein axils, puberulous below along principal veins, often with simple and dendroid trichomes in vein axils, plate-shaped glands below mostly in axils of secondary veins but also scattered along midrib; tip acuminate; base cuneate to rounded; margins entire. Tendrils simple. Inflorescence a raceme, axillary or terminal. Calyx yellow, 2–6 cm long, 1.2–4.1 cm wide, glabrous (sometimes stellate or simple puberulent near base) to rarely stellate-pubescent, with scattered plate-shaped glands; corolla tubular-campanulate, 5.5–11 cm long; corolla tube 4.5–7+ cm long. Corolla lobes 1–3+ cm long, yellow; throat sometimes reddish internally, sparsely and often inconspicuously lepidote internally and externally. Fruit a capsule, 22–32 cm long, 6–12 cm wide, ellipsoid, flattened, glabrous, somewhat wrinkled. Seeds winged, 2.5–4+ cm long, 6–13 cm wide.
Phenology. Flowers from February through October (wet season). Fruits from January through March (dry season).
Distribution. Mexico to Brazil. In Costa Rica found from 100 to 1,000 m.
Notes. Called *flor de guineo* in Nicaragua.

5. GENUS *CYDISTA*

Cydista is one of only two genera in our area with a garlic or onion odor to the vegetative parts. The other genus, *Mansoa*, is readily separated by the trifid tendrils (simple in *Cydista*).

1. *Cydista aequinoctialis* (L.) Miers, Proc. Roy. Hort. Soc. London 3: 191. 1863. Figure 3.3A.

Synonymy. *Bignonia aequinoctialis* L., Sp. pl. ed. 1. 2: 623. 1753; *B. spectabilis* Vahl, Symb. bot. 80. 1794; *B. sarmentosa* Bertol., Fl. Guatimal. 25. 1840; *B. sarmentosa* Bertol. var.

hirtella Benth., Bot. voy. Sulphur 128. 1844; *Cydista sarmentosa* (Bertol.) Miers, Proc. Roy. Hort. Soc. London 3: 191. 1863; *C. spectabilis* (Vahl) Miers, Proc. Roy. Hort. Soc. London 3: 191. 1863; *Levya nicaraguensis* Bureau ex Baill., Hist. Pl. 10: 29. 1888; *Arrabidaea guatemalensis* K. Schum. & Loes., Bot. Jahrb. Syst. 23: 129. 1896; *A. pseudochica* Kraenzl., Feddes Repert. 17: 19. 1921; *Anemopaegma tonduzianum* Kraenzl., Feddes Repert. 17: 116. 1921; *Cydista pubescens* S.F. Blake, Contr. U.S. Natl. Herb. 24: 23. 1922; *Arrabidaea isthmica* Standl., J. Wash. Acad. Sci. 15: 461. 1925, pro parte.

Description. Liana, sometimes with onion or garlic odor to vegetative parts; stems with eight phloem arms in cross-section; branchlets subtetragonal to tetragonal, glabrous to pubescent; interpetiolar gland fields absent. Leaves bi- or trifoliolate; terminal leaflet sometimes replaced by a tendril; leaflets more or less ovate, to 16 cm long and 9 cm wide, lepidote above and below, variously pubescent below, glandular fields in many secondary vein axils; tip acuminate to obtuse; base cuneate to truncate; margins entire. Tendrils simple. Inflorescence a panicle, axillary or terminal. Calyx cupular, 4–10 mm long and wide, lepidote to pubescent, glandular externally. Corolla tubular-campanulate, to 7.5 cm long; tube to 5 cm long, white to lavender, sometimes with maroon lines at base of lobes and a yellow throat opening; lobes puberulous internally, glandular-lepidote externally, to 2.5 cm long. Fruit a capsule, 20–40 cm long, 1.5–2.5 cm wide, linear, flattened, lepidote, surface somewhat wrinkled. Seeds winged, to 1.8 cm long, 7+ cm wide.

Phenology. Flowers from January to September. Fruits from January to March, and July to August.

Distribution. West Indies, Mexico to Brazil. In Costa Rica found from sea level to 1,100 m.

Notes. In El Salvador this species is commonly called *ajillo* and *bejuco de ajo*, referring to the garlic or onion odor of freshly crushed leaves (Standley and Calderon 1925). In Costa Rica it has also been called *jalapa*. Mayan names compiled by Gaumer and Schott (Standley 1930) include *chacnetoloc*, *anicab*, and *zolak* (or *tsolak*).

A pubescent form of this species has been called *Cydista sarmentosa* or *Cydista pubescens*. However, Gentry (1973) suggested that the form should rather be called *Cydista aequinoctialis* var. *hirtella* (Benth.) A.H. Gentry. In Nicaragua the pubescent form is commonly called *jalapa* or *bejuco blanco*.

6. GENUS *DISTICTELLA*

Only one species of *Distictella* is found in our area, *Distictella magnoliifolia*. Gentry (1973) noted that the trifid tendrils and the subterete, smooth, reddish, thick branchlets were useful field characters.

1. ***Distictella magnoliifolia*** (Kunth) Sandwith, Lilloa 3: 460. 1938. Figure 3.3B.

Synonymy. *Bignonia magnoliaefolia* [sic] Kunth in H.B.K., Nov. Gen. sp. pl. 3: 136. 1819; *Distictella kochii* (Pilg.) Urb., Feddes Repert. 14: 310. 1916; *D. racemosa* (Bureau & K. Schum.) Urb., Feddes Repert. 14: 310. 1916.

Description. Liana; stems with four phloem arms in cross-section (at least when young); branchlets subterete to terete, puberulous; interpetiolar gland fields absent. Leaves bi- or trifoliolate; terminal leaflet sometimes replaced by a tendril; leaflets elliptic, to 25 cm long and 12 cm wide, somewhat lepidote above and more so below, puberulous below, pubescent in axils of secondary veins, sometimes with scattered glands near the midrib; tip acute to acuminate; base cuneate to truncate; margins entire. Tendrils trifid. Inflorescence a racemose panicle. Calyx cupular, to 10 mm long and slightly less wide, puberulous with

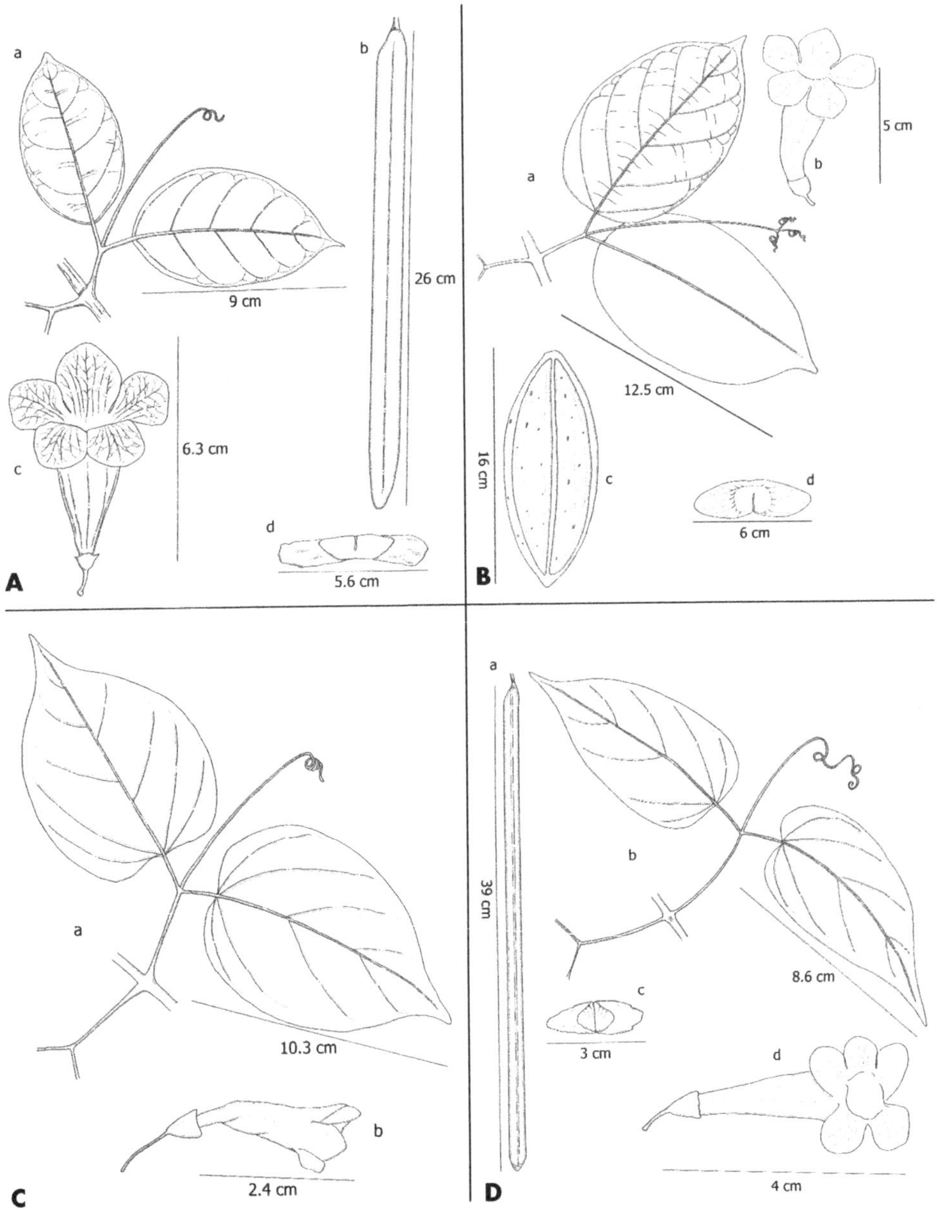

Fig. 3.3. A. *Cydista aequinoctialis:* **(a)** leaf and tendril, **(b)** fruit, **(c)** flower, **(d)** seed (Leaf, tendril, and flower based on MacDougal 1146, DUKE; Fruit and seed based on Moreno 9870, MO). **B.** *Distictella magnoliifolia:* **(a)** leaf and tendril, **(b)** flower, **(c)** fruit, **(d)** seed (Leaf and flower based on Smith et al. 1628, US; Fruit and seed based on Solomon 18428, MO). **C.** *Lundia corymbifera:* **(a)** leaf and tendril, **(b)** flower (Based on Wilbur 21092, F). **D.** *Lundia puberula:* **(a)** fruit, **(b)** leaf and tendril, **(c)** seed, **(d)** flower (Leaf and tendril based on Steyermark 38814, F; Flower based on Atwood and Narvaez 3234, F; Fruit and seed based on Gentry 7404, MO).

linear gland fields near upper half. Corolla tubular-campanulate, to 6 cm long; tube to 4 cm long, white, glabrous internally except for glandular trichomes near point of stamen insertion, pubescent externally; lobes puberulous externally, to 1.2 cm long. Fruit a capsule, to 15 cm long and 6 cm wide, ellipsoid, somewhat flattened, tomentose, with scattered glandular areas, midrib inconspicuous. Seeds to 2.3 cm long and 4 cm wide.

Phenology. Flowering times unknown. Fruits in November.

Distribution. Costa Rica to Brazil. In Costa Rica found from sea level to 700 m.

7. GENUS *LUNDIA*

Two species of *Lundia* occur in our area. The genus is characterized by interpetiolar gland fields, terete or subterete branchlets, and stems with four phloem arms in cross-section. It is distinguished from the similar-looking *Arrabidaea* by pubescent anthers (glabrous in *Arrabidaea*), and pubescent to velutinous capsules (glabrous or with scattered trichomes to tuberculate in *Arrabidaea*). Other vegetative characteristics include blunt pseudostipules and broadly ovate to cordate, somewhat ciliate leaves.

KEY TO *LUNDIA*

1. Corolla white; style glabrous (sometimes puberulous at extreme base)... *L. corymbifera*, p. 25.
1. Corolla pink to lavender or magenta; style puberulous throughout or only at apex... *L. puberula*, p. 25.

1. ***Lundia corymbifera*** (Vahl) Sandwith, Recueil Trav. Bot. Néerl. 34: 229. 1937. Figure 3.3C.

Synonymy. *Bignonia corymbifera* Vahl, Eclog. Amer. 2: 45, pl. 17. 1798; *Lundia umbrosa* (Kunth) Bureau, Adansonia 8: 282. 1868; *Arrabidaea corymbifera* (Vahl) Bureau ex K. Schum. in Engler & Prantl, Nat. Pflanzenfam. 4: 213. 1894; *Cuspidaria corymbifera* (Vahl) Baill. ex K. Schum. in Engler & Prantl, Nat. Pflanzenfam. 4: 216. 1894.

Description. Liana; stems with four phloem arms in cross-section; branchlets terete, puberulous; interpetiolar gland fields present. Leaves bi- or trifoliolate; terminal leaflet sometimes replaced by a tendril; leaflets more or less ovate, to 10 cm long and 7 cm wide, puberulous above and below at least in secondary vein axils, scattered glands below near midrib; tip acuminate; base truncate to obliquely cordate; margins entire. Tendrils simple. Inflorescence a panicle, axillary or terminal. Calyx finely puberulous, 3–6 mm long, more or less 4.5 mm wide. Corolla white, to 4.5 cm long; tube puberulous externally, to nearly 3 cm long; lobes glandular-lepidote and puberulous internally, puberulous externally, to about 1.4 cm long. Fruit a capsule, to 45 cm long, about 1.6 cm wide, linear, somewhat flattened, velutinous; midvein conspicuously raised. Seeds winged, to 1.2 cm long and 3.5 cm wide.

Phenology. Flowers from June through August. Fruits in March.

Distribution. Costa Rica to Bolivia. In Costa Rica found from sea level to 900 m.

2. ***Lundia puberula*** Pittier, Contr. U.S. Natl. Herb. 18: 258. 1917. Figure 3.3D.

Synonymy. *L. dicheilocalyx* Blake, Contr. Gray Herb. 52: 94. 1917; *L. schumanniana* Kraenzl., Feddes Repert. 17: 120. 1921.

Description. Liana; stems with four phloem arms in cross-section; branchlets terete, puberulous; interpetiolar gland fields present. Leaves bi- or trifoliolate; terminal leaflet sometimes replaced by a tendril; leaflets more or less ovate, to 9 cm long and nearly 6 cm wide, puberulous above and below, sometimes with longer trichomes below in secondary vein axils, scattered glands below near midrib; tip acuminate; base truncate to obliquely cordate; margins entire. Tendrils simple. Inflorescence a panicle, axillary or terminal. Calyx finely puberulous, 4–6 mm long, more or less 3.5 mm wide. Corolla tubular-campanulate, to 5.6 cm long, pink to lavender or magenta; tube puberulous externally, to 4 cm long; lobes glandular-lepidote and puberulous internally, puberulous externally, to about 2 cm long. Fruit a capsule, to 45 cm long, about 1.6 cm wide, linear, somewhat flattened, velutinous; midvein conspicuously raised. Seeds winged, to 1.2 cm long and 3.5 cm wide.

Phenology. Flowers from February through May and from September through December. Fruiting times unknown.

Distribution. Belize and southern Mexico to Venezuela. In Costa Rica found from 100 to 1,100 m.

8. GENUS *MACFADYENA*

Macfadyena is one of only two or three genera in our area with uncinate tendrils. In the second genus, *Melloa*, the tendrils frequently become woody with time, whereas in *Macfadyena*, they remain herbaceous. The third genus, *Tynanthus*, may also have uncinate tendrils, but it is too poorly known to be certain. Juvenile plants of *Macfadyena* (usually unmistakable in their greatly reduced leaf size) also frequently root adventitiously at the nodes. Both *Macfadyena* and *Melloa* are possibly our only genera with irregularly many-armed phloem in cross-sections of stem. The stem cross-section of *Tynanthus* is not known. *Tynanthus* can be separated from both other genera by the tannish pubescence on abaxial leaf surfaces. Leaves of both other genera are more or less glabrous on both surfaces or only somewhat puberulent on the principal veins.

1. ***Macfadyena unguis-cati*** (L.) A.H. Gentry, Brittonia 25: 236. 1973. Figure 3.4A.

Synonymy. *Bignonia unguis-cati* L., Sp. pl., ed. 1. 2: 623. 1753.

Description. Liana; stems with phloem irregularly many-armed in cross-section, juveniles rooting adventitiously at nodes. Plant dimorphic; branchlets terete, glabrous to puberulous; interpetiolar gland fields usually absent. Leaves of mature plants bi- or trifoliolate, to 16 cm long and 7 cm wide; terminal leaflets sometimes replaced by a tendril; leaflets more or less ovate, 2 cm long, 0.8 cm wide, scattered lepidote above and below, otherwise glabrous above, sometimes puberulous below along veins, scattered plate-shaped glands below near midrib; tip acute to acuminate; base cuneate to truncate; margins entire. Tendrils trifid, uncinate, replacing terminal leaflet. Inflorescence a panicle, axillary. Calyx cupular, 5–18 mm long and about equally wide, glabrous to somewhat lepidote, frequently with scattered glands. Corolla tubular-campanulate, to 10 cm long, yellow, sometimes with orange lines in throat; tube glabrous to puberulous internally except pubescent near point of stamen insertion, glabrous externally, nearly 7 cm long; lobes to 3 cm long. Fruit a capsule, to 100 cm long and about 1.8 cm wide, linear, flattened, inconspicuously lepidote. Seeds winged, to 1.8 cm long, 5.5+ cm wide.

Phenology. Flowers from January through July. Fruits from April through June.

Distribution. Mexico to Argentina. In Costa Rica found from sea level to 840 m.

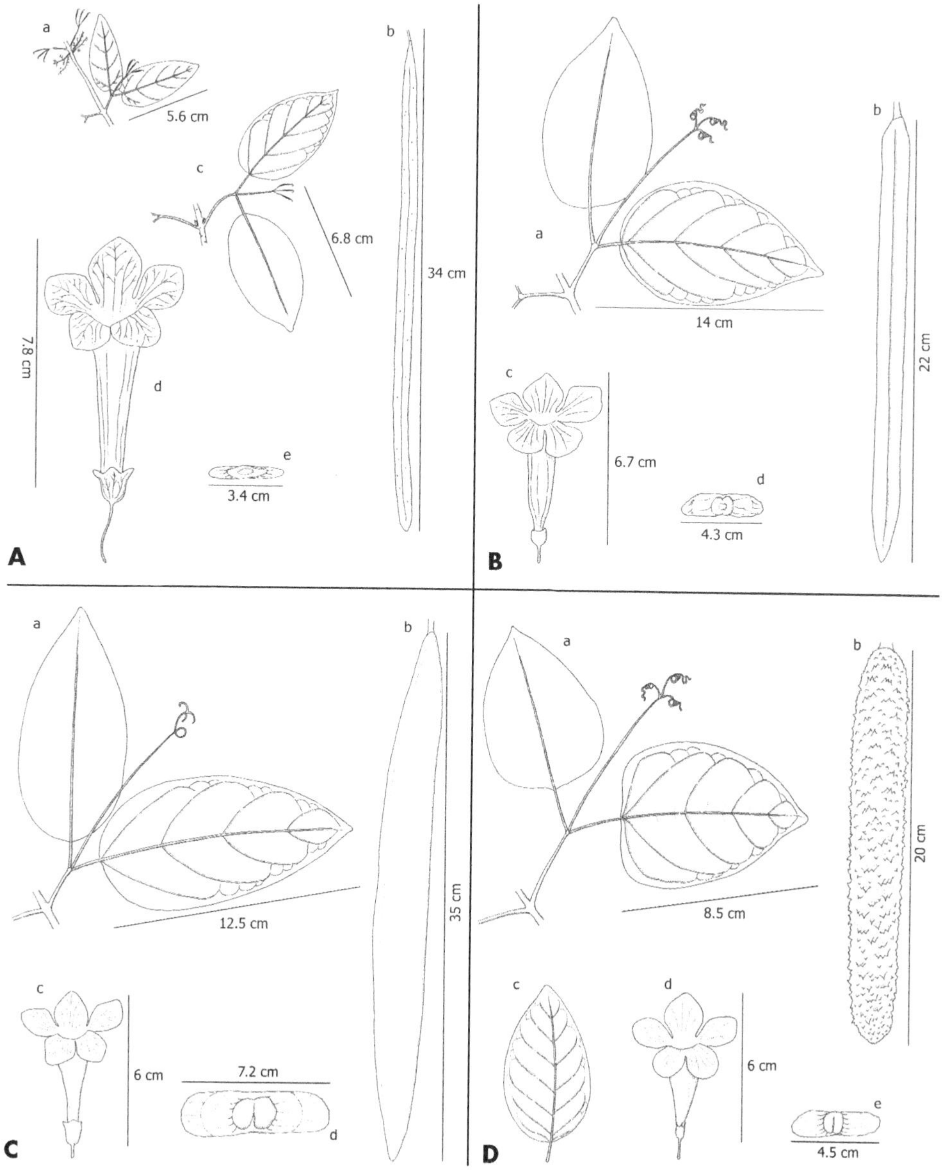

Fig. 3.4. A. *Macfadyena unguis-cati:* **(a)** leaf, tendrils, and adventitious roots of juvenile form; **(b)** fruit; **(c)** leaf and tendril of mature form; **(d)** flower; **(e)** seed (Leaves and tendrils of both forms and adventitious roots of juvenile form based on Standley 40203, US; Flower based on Breedlove 50335, DUKE; Fruit and seed based on Matuda 18373, MO). **B.** *Mansoa hymenaea:* **(a)** leaf and tendril, **(b)** fruit, **(c)** flower, **(d)** seed (Leaf based on Glaziou 4690, US; Flower based on Woytkowski 6793, US; Fruit and seed based on Morales 2558, MO). **C.** *Mansoa standleyi:* **(a)** leaf and tendril, **(b)** fruit, **(c)** flower, **(d)** seed (Leaf and tendril based on Schunke 2806, US; Flower based on Klug 4138, US; Fruit and seed based on Schunke 14262, MO). **D.** *Mansoa verrucifera:* **(a)** leaf and tendril, **(b)** fruit, **(c)** leaflet variation, **(d)** flower, **(e)** seed (Leaves and tendril based on Dugand 5676, US; Flower based on Peterson J-112, US; Fruit and seed based on Grayum et al. 4964, MO).

Notes. This species is sometimes known as *uña de gato*, a name also frequently applied to *Uncaria tomentosa* (Willd. ex Roem. & Schult) DC. (Rubiaceae).

9. GENUS *MANSOA*

Mansoa is one of only two genera in our area that usually produces a garlic or onion odor from bruised vegetative parts. The other genus, *Cydista*, is usually distinguished by simple tendrils (trifid in our *Mansoa*). In Costa Rica only *M. parvifolia* has simple tendrils, but it is a low-elevation species. *Mansoa parvifolia* has neither an onion or garlic odor nor interpetiolar gland fields and can additionally be distinguished from all other bignoniaceous species in Costa Rica by having disc-tipped tendrils. Other *Mansoa* have interpetiolar gland fields, except for *M. verrucifera* (distinguished by its strongly tuberculate capsule).

KEY TO *MANSOA*

1. Interpetiolar gland fields lacking; capsule tuberculate... *M. verrucifera*, p. 29.
1. Interpetiolar gland fields present; capsule not tuberculate
   2. Calyx shorter than 1 cm; gland fields present or absent below in axils of basal veins of leaflets; fruits slightly compressed, median nerve conspicuously raised... *M. hymenaea*, p. 28.
   2. Calyx longer than 1.3 cm; gland fields present below in axils of basal veins of leaflets; fruits strongly compressed; median nerve mostly obscure... *M. standleyi*, p. 29.

1. *Mansoa hymenaea* (DC.) A.H. Gentry, Ann. Missouri Bot. Gard. 66: 782. 1979. Figure 3.4B.

Synonymy. *Bignonia hymenaea* DC., Prodr. 9: 159. 1845; *Adenocalymma macrocarpum* Donn. Sm., Bot. Gaz. 40: 9. 1915; *A. obovatum* Urb., Feddes Repert. 14: 307. 1916; *A. ciliolatum* Blake, Contr. Gray Herb. 52: 90. 1917; *A. hosmeca* Pittier, Contr. U.S. Natl. Herb. 18: 256. 1917; *Petastoma landglasseanum* Kraenzl., Feddes Repert. 17: 56. 1921; *P. tonduzianum* Kraenzl., Feddes Repert. 17: 56. 1921; *Adenocalymma alboviolaceum* Loes., Verh. Bot. Vereins Prov. Brandenburg 65: 100. 1923; *Pseudocalymma hymenaeum* (DC.) Sandwith, Candollea 7: 247. 1937; *P. langlasseanum* (Kraenzl.) Sandwith, Recueil Trav. Bot. Néerl. 34: 210. 1937; *P. macrocarpum* (Donn. Sm.) Sandwith, Recueil Trav. Bot. Néerl. 34: 210. 1937; *P. pachypus* (K. Schum.) Sandwith, Recueil Trav. Bot. Néerl. 34: 210. 1937; *P. pohlianum* (Bureau & K. Schum.) Sandwith, Recueil Trav. Bot. Néerl. 34: 211. 1937; *P. alliaceum* var. *microcalyx* Sandwith, Kew Bull. 1953: 467. 1954, pro parte, typo excluso.

Description. Liana; strong garlic odor to all vegetative parts; stems with four to eight phloem arms in cross-section; branchlets subterete, glabrous to somewhat lepidote; interpetiolar gland fields present. Leaves opposite, bi- or trifoliolate; terminal leaflet sometimes replaced by a tendril; leaflets more or less ovate, to 9 cm long and nearly 7 cm wide, glabrous above and below, sometimes basally lepidote, frequently with gland field below in basal secondary vein axil; tip acuminate; base truncate, subcordate or rounded; margins entire. Tendrils trifid. Inflorescence a panicle, terminal. Calyx cupular, 4–6 mm long and wide, finely puberulous to lepidote. Corolla tubular-campanulate, to 5.8 cm long, whitish to lavender or magenta; tube glabrous internally, to 4 cm long; lobes puberulous, to about 1.8 cm long. Fruit a capsule, to 23 cm long, about 1.8 cm wide, linear, flattened, glabrous to somewhat lepidote; midvein conspicuously raised. Seeds winged, to 1.5 cm long and 3.5 cm wide.

Phenology. Flowers from December through July. Fruits in May.

Distribution. Mexico to Brazil. In Costa Rica found from sea level to 900 m.

2. *Mansoa standleyi* (Steyerm.) A. H. Gentry, Ann. Missouri Bot. Gard. 66: 783. 1979. Figure 3.4C.

Synonymy. *Pseudocalymma standleyi* Steyerm., Publ. Field Mus. Nat. Hist., Bot. Ser., 23: 235. 1947.

Description. Liana; strong garlic odor to all parts; stems with eight phloem arms in cross-section; branchlets subterete to subtetragonal, glabrous to somewhat lepidote; interpetiolar gland fields present. Leaves bi- (or tri-) foliolate; terminal leaflet sometimes replaced by a tendril; leaflets more or less elliptic, 8–20 cm long, 4–10 cm wide, glabrous above and below, sometimes lepidote basally, frequently with gland field below in basal secondary vein axil; tip acuminate; base cuneate to rounded; margins entire. Tendrils trifid. Inflorescence a panicle, axillary or terminal. Calyx cupular, 9–21 mm long, 8–16 mm wide, glabrous, sometimes with raised glands. Corolla tubular-campanulate, to 8 cm long; base yellow; apex and lobes reddish, yellow to pinkish internally; tube glabrous internally and externally, except for trichomes near point of stamen insertion, to 6 cm long; lobes puberulous, to about 2.8 cm long. Fruit a capsule, to 65+ cm long, about 3.8 cm wide, linear, flattened, glabrous; midvein inconspicuous. Seeds winged, to 2.8 cm long and 11 cm wide.

Phenology. Flowers in March and April. Fruiting season unknown.

Distribution. Guatemala to Venezuela and Peru. In Costa Rica found from sea level to 1,100 m.

3. *Mansoa verrucifera* (Schltdl.) A.H. Gentry, Ann. Missouri Bot. Gard. 63: 62. 1976. Figure 3.4D.

Synonymy. *Bignonia verrucifera* Schltdl., Linnaea 26: 655. 1853; *Onohualcoa verrucifera* (Schltdl.) A.H. Gentry, Ann. Missouri Bot. Gard. 60: 885. 1974.

Description. Liana; garlic odor present in some vegetative parts or the wood; stems with eight phloem arms in cross-section; branchlets subterete, glabrous; interpetiolar gland fields absent. Leaves bi- or trifoliolate; terminal leaflet sometimes replaced by a tendril; leaflets more or less ovate, to 13 cm long and 6 cm wide, puberulous to glabrous above and below, frequently with gland field below in basal secondary vein axils; tip acute; base cuneate to subcordate; margins entire. Tendrils trifid. Inflorescence a panicle, terminal. Calyx cupular, 5–7 mm long and wide, glandular. Corolla tubular-campanulate, to 8 cm long, whitish to lavender or pinkish; throat often white; tube puberulous, to 6 cm long, glabrous internally, glandular lepidote externally; lobes to about 2.5 cm long. Fruit a capsule, to 30 cm long, about 3+ cm wide, linear-oblong, strongly tuberculate. Seeds winged, to 2.5 cm long and 7.5 cm wide.

Phenology. Flowers in March. Fruits from January to April.

Distribution. Southern Mexico to British Guiana (Surinam). In Costa Rica found from 300 to 1,000 m.

10. GENUS *MARTINELLA*

In fruit, our only species in this genus, *Martinella obovata*, is almost unmistakable. Although linear, flattened capsules are common in the Bignoniaceae, none grow as long as those of *Martinella* (to 130+ cm). Dependable vegetative characters include long, frequently bending or twisting petiolules, trifid tendrils, and conspicuous interpetiolar swellings (Gentry 1973).

1. *Martinella obovata* (Kunth) Bureau & K. Schum. in Mart., Fl. Bras. 8: 161, pl. 84. 1896. Figure 3.5A.

Synonymy. *Spathodea obovata* Kunth in H.B.K., Nov. Gen. sp. pl. 3: 147. 1819; *Bignonia obovata* (Kunth) Spreng., Syst. veg. 2: 830. 1825; *Tabebuia cordata* Benth., Bot. voy. Sulphur 129. 1844; *Periarrabidaea duckei* (Samp.) A. Samp., Anais Acad. Brasil. Ci. 12: 91. 1936.

Description. Liana; stems with four phloem arms in cross-section; branchlets subterete to terete, glabrous to puberulous; interpetiolar gland fields absent; interpetiolar ridge present. Leaves bi- or trifoliolate; terminal leaflet sometimes replaced by a tendril; leaflets more or less ovate or elliptic, to 20 cm long and 16 cm wide, puberulous above and below at least on principal veins, frequently with glands below scattered near midrib; tip acuminate; base truncate to subcordate; margins entire. Tendrils trifid. Inflorescence a raceme, axillary. Calyx tubular-campanulate, 1–2 cm long, 0.7–1.2 cm wide, puberulous, sometimes with glandular trichomes. Corolla tubular-campanulate, to 6 cm long, deep violet magenta to deep red-wine; tube glabrous to glandular-lepidote externally, to 5 cm long; lobes to about 1.5 cm long. Fruit a capsule, to 130+ cm long, about 1.8 cm wide, linear, flattened, glabrous to somewhat puberulous. Seeds winged, to 1.8 cm long and 5+ cm wide.

Phenology. Flowers from February through March. Fruits from April through November, and in January.

Distribution. Belize to Bolivia. In Costa Rica found from 300 to 1,450 m.

11. GENUS *MELLOA*

Melloa is one of only two genera in Costa Rica definitely known to have uncinate tendrils. It can be separated from *Macfadyena* because the tendrils become woody with age (but remain herbaceous in *Macfadyena*). Also, juvenile *Melloa* does not root adventitiously at the nodes as in *Macfadyena*. A third genus, *Tynanthus*, may also be uncinate, but is too poorly known to be certain. *Tynanthus* can be separated from both other genera by the tannish pubescence on leaf undersurfaces. Both *Melloa* and *Macfadyena* are our only genera where the phloem is irregularly many-armed in cross-section. The cross-section of the phloem of *Tynanthus* is unknown.

1. *Melloa quadrivalvis* (Jacq.) A.H. Gentry, Brittonia 25: 237. 1973. Figure 3.5B.

Synonymy. *Bignonia quadrivalvis* Jacq., Fragm. bot. 37, tab. 40, fig. 3. 1800–1809; *Phryganocydia pisoniana* Mart. ex DC., Prodr. 9: 209. 1845; *Tabebuia pisoniana* (DC.) Miers, Proc. Roy. Hort. Soc. London 3: 199. 1863; *Tecoma pisoniana* DC. ex Bureau & K. Schum. in Mart., Fl. Bras. 8: 312. 1897; *Melloa populifolia* (DC.) Britton, Ann. New York Acad. Sci. 7: 188. 1893.

Description. Liana; stems with phloem irregularly many-armed in cross-section; branchlets subterete to terete, glabrous to puberulous; interpetiolar gland fields absent; interpetiolar

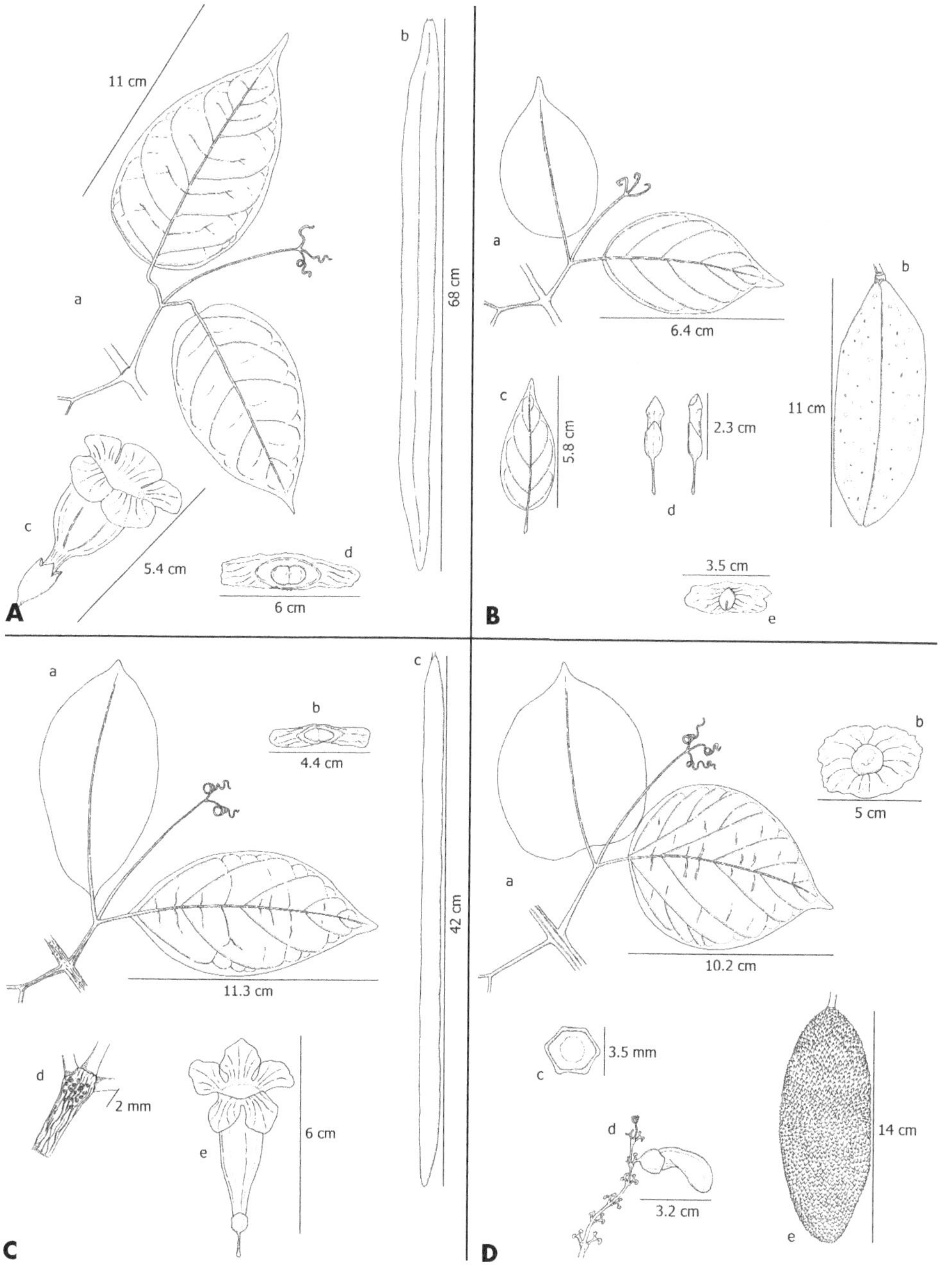

Fig. 3.5. A. *Martinella obovata:* **(a)** leaf and tendril, **(b)** fruit, **(c)** flower, **(d)** seed (Leaf, tendril, and flower based on Wilbur 37662, DUKE; Fruit and seed based on Chacón 400, MO). **B.** *Melloa quadrivalvis:* **(a)** leaf and tendril, **(b)** fruit, **(c)** leaflet variation, **(d)** flower buds, **(e)** seed (Leaf based on Moreno 21780, DUKE; Leaflet variation, tendril, and flower buds based on Duke M3629, DUKE; Fruit and seed based on Bello 3620, MO). **C.** *Paragonia pyramidata:* **(a)** leaf and tendril, **(b)** seed, **(c)** fruit, **(d)** close-up of petiole apex showing gland field, **(e)** flower (Leaf, tendril, petiole apex, and flower based on Jacobs 3083, DUKE; Fruit and seed based on Balick 2218, MO). **D.** *Pithecoctenium crucigerum:* **(a)** leaf and tendril, **(b)** seed, **(c)** stem cross-section, **(d)** inflorescence, **(e)** fruit (Leaf, stem cross-section, and inflorescence based on Foster 840, DUKE; Fruit and seed based on Fryxell and Lott 3248, MO).

ridge frequently present. Leaves bi- or trifoliolate; terminal leaflet sometimes replaced by a tendril; leaflets more or less elliptic, to 10 cm long and 5 cm wide, glabrous above and below, scattered lepidote below; tip acuminate; base truncate to subcordate; margins entire. Tendrils trifid, uncinate. Inflorescence a panicle, axillary or terminal. Calyx campanulate, 1–2 cm long, 0.9–1.2 cm wide, glabrous. Corolla tubular-campanulate, to 7 cm long, yellow; tube pubescent internally, glabrous externally, to 5 cm long; lobes somewhat ciliate, to about 1.5 cm long. Fruit a capsule, to 15 cm long, about 5+ cm wide, oblong, minutely tuberculate. Seeds winged, to 1.6 cm long and 5+ cm wide.

Phenology. Flowers from February through May and in August. Fruits in March and from August through December.

Distribution. Mexico to Argentina. In Costa Rica found from 50 to 1,100 m.

12. GENUS *PARAGONIA*

The only *Paragonia* of the continental Neotropics, *P. pyramidata*, is recognized by bifid or trifid tendrils, stems with four phloem arms in cross-section, and glandular fields on the top sides of the tip of the petiole.

1. *Paragonia pyramidata* (Rich.) Bureau, Vidensk. Meddel. Dansk Naturhist. Foren. Kjøbenhavn 1893: 104. 1894. Figure 3.5C.

Synonymy. *Bignonia pyramidata* Rich., Actes Soc. Hist. Nat. Paris 1: 110. 1792; *B. striata* DC., Rev. Bign. (Bibl. Univ. Geneve) 22. 1838, nomen nudum; *B. sinclairii* Benth., Bot. voy. Sulphur 129. 1844; *Tabebuia pyramidata* (Rich.) DC., Prodr. 9: 214. 1845; *Temnocydia elliptica* Mart. ex DC., Prodr. 9: 156. 1845, pro syn.; *T. lenta* Mart. ex DC., Prodr. 9: 159. 1845, pro syn.; *Arrabidaea dichasia* Donn. Sm., Bot. Gaz. 20: 6. 1895.

Description. Liana; stems with four phloem arms in cross-section; branchlets terete, glabrous or puberulous when young, somewhat lepidote; interpetiolar gland fields usually absent; interpetiolar ridge frequently present. Leaves bi- or trifoliolate; terminal leaflet sometimes replaced by a tendril; leaflets more or less ovate or elliptic, to 20 cm long and 12 cm wide, puberulous below at least on midrib, frequently with glands below scattered near midrib; tip acute to short acuminate; base cuneate to rounded; margins entire. Tendrils usually simple or bifid, sometimes trifid. Inflorescence a panicle, terminal. Calyx cupular, 0.4–0.7 cm long, 4–10 cm wide, lepidote. Corolla tubular-campanulate, to 5.5 cm long, magenta or lavender but tip and base of tube often white; tube glabrous internally, to 4 cm long; lobes pubescent, to about 1.5 cm long. Fruit a capsule, to 60 cm long, about 1.5 cm wide, linear, convex, finely tuberculate. Seeds winged, to 1.3 cm long, 4+ cm wide.

Phenology. Flowers from January through July. Fruits from March through July and in November.

Distribution. Mexico to Bolivia. In Costa Rica found from sea level to 1,200 m.

13. GENUS *PITHECOCTENIUM*

Pithecoctenium is readily recognized by its regularly hexagonal stems. It is distinguished from the similar-looking *Amphilophium* (also with cordate leaflet bases and hexagonal stems) by simple trichomes (dendroid or forked in *Amphilophium*). The ornamental *Pyrostegia* differs by its orange flowers (white in *Pithecoctenium* and yellowish-white to purple in *Amphilophium*) and leaflets with rounded or truncate bases.

1. *Pithecoctenium crucigerum* (L.) A.H. Gentry, Taxon 24: 123. 1975. Figure 3.5D.

Synonymy. *Bignonia crucigera* L., Sp. pl. 623. 1753; *Pithecoctenium echinatum* (Jacq.) Baill., Hist. Pl. 10: 8. 1888.

Description. Liana; stems with four phloem arms in cross-section; branchlets sharply hexagonal, angles ribbed, lepidote with scattered trichomes; interpetiolar gland fields absent; interpetiolar ridge sometimes present. Leaves bi- or trifoliolate; terminal leaflet sometimes replaced by a tendril; leaflets more or less ovate to suborbicular, to 18 cm long and 15 cm wide, variously pubescent above and below, frequently with plate-shaped glands below scattered in axils of veins; tip acuminate; base truncate to cordate; margins entire. Tendrils trifid to many-branching. Inflorescence a raceme, terminal. Calyx cupular, to 1.2 cm long and 1 cm wide, pubescent, eglandular. Corolla tubular-campanulate, frequently bent at 90 degrees in middle, to 6 cm long, yellow with white throat; tube to 5.2 cm long; lobes pubescent externally, to about 1.7 cm long. Fruit a capsule, to 30 cm long, about 7+ cm wide, oblong-elliptic, echinate-tuberculate. Seeds winged, to 4 cm long, 9+ cm wide.

Phenology. Flowers from April through June and from October through November. Fruits from February to March, June to July, and October to December.

Distribution. Mexico to Argentina. In Costa Rica found from sea level to 1,100 m.

Notes. In Costa Rica and Panama this species is sometimes known as *peine de mono* (monkey's comb). Mayan names include *xachextabay*, *netoloc* (apparently iguana tail), and *xtabay* (Standley 1930). In Yucatan the plant may also be called *peine de mico*.

14. GENUS *PLEONOTOMA*

Our only member of this genus, *Pleonotoma variabilis*, is recognized by the bi- or triternate leaves and entire leaflet margins. The only other bignoniaceous taxon with bi- or triternate leaves is *Tourrettia* (herbaceous with strongly serrated leaflet margins).

1. *Pleonotoma variabilis* (Jacq.) Miers, Proc. Roy. Hort. Soc. London 3: 184. 1963. Figure 3.6A.

Synonymy. *Bignonia variabilis* Jacq., Pl. hort. schoenbr. 2: 45, *tab.* 212. 1797.

Description. Liana; stems with four phloem arms in cross-section; branchlets sharply tetragonal, angles ribbed; interpetiolar gland fields absent; interpetiolar ridge present. Leaves bi- or triternate; terminal leaflet sometimes replaced by a tendril; leaflets more or less ovate or elliptic, to 16 cm long and 9 cm wide, puberulous above at least on midrib, trichomes below in axils of secondary veins; tip acute or acuminate; base obliquely truncate or cuneate; margins entire. Tendrils trifid, replacing terminal leaflet. Inflorescence a raceme, usually terminal. Calyx cupular, 5–9 mm long and about equally wide, glabrous, sometimes ciliate with scattered lepidote scales and glands near apex. Corolla white or cream, to 10 cm long; tube yellowish, to 7 cm long, glabrous internally and externally (except pubescent near point of stamen insertion); lobes glandular-lepidote, to 2+ cm long, puberulous. Fruit a capsule, to 30 cm long, about 2+ cm wide, linear-oblong, somewhat compressed, warty-tuberculate. Seeds winged, to 1.5 cm long and 5 cm wide.

Phenology. Flowers from March through May and in December. Fruits in September.

Distribution. Costa Rica to Venezuela. In Costa Rica found from sea level to 700 m.

15. GENUS *PYROSTEGIA*

Frequently grown as an ornamental in Costa Rica, the orange flowers of *Pyrostegia venusta* are commonly seen throughout the year along garden walls and courtyards. Vegetatively the species is recognized by its irregularly to regularly 6–8 angled or ribbed stems, and by the rounded to truncate leaflet bases.

1. ***Pyrostegia venusta*** (Ker.) Miers, Proc. Roy. Hort. Soc. London 3: 188. 1863. Figure 3.6B.

Synonymy. *Bignonia venusta* Ker Gawl, Bot. Reg. 3: 5. 249. 1818; *B. ignea* Vell., Fl. flumin. 244. 1825; *Pyrostegia ignea* (Vell.) C. Presl, Bot. Bemerk. 93. 1843.

Description. Liana; branchlets irregularly or regularly 6–8 angled (generally not tetragonal), puberulous; interpetiolar gland fields absent. Leaves bifoliolate, sometimes with a tendril; leaflets more or less ovate, to 6 cm long and 4 cm wide, puberulous or lepidote; tip acute or acuminate; base truncate to rounded (but not cordate); margins entire. Tendrils trifid. Inflorescence a panicle, terminal. Calyx cupular, about 8 mm long, 4 mm wide, lepidote; margins ciliate. Corolla tubular, to 6.5 cm long, red or orange; tube pubescent internally below point of stamen insertion, glabrous externally, to 5 cm long; lobes to about 1.5 cm long. Fruit a capsule, to 30 cm long, about 1.5 cm wide, linear; midrib slightly raised. Seeds winged, to 1.4 cm long, 4+ cm wide.

Phenology. Flowers and fruits throughout the year.

Distribution. Believed to be native to Brazil, Paraguay, and Bolivia. From frequent cultivation and possible naturalization, it is now found throughout Central America and Mexico.

Notes. This species is principally found in gardens in Costa Rica.

16. GENUS *TOURRETTIA*

Tourrettia is our only herbaceous, tendrillate member of the Bignoniaceae. It is further characterized by biternate leaves and strongly serrate leaflet margins. The only other species of Bignoniaceae with bi- or triternate leaves is *Pleonotoma variabilis*. The two may be distinguished by the leaflet margins (serrate in *Tourrettia* and entire in *Pleonotoma*), habit (herbaceous vine in *Tourrettia* and woody liana in *Pleonotoma*), and capsule surface and size (densely echinate and to 4.5 cm long in *Tourrettia*; not echinate and longer than 10 cm in *Pleonotoma*).

1. ***Tourrettia lappacea*** (L'Hér.) Willd. in L., Sp. pl., ed. 4. 3: 263. 1801. Figure 3.6C.

Synonymy. *Dombeya lappacea* L'Hér., Stirp. nov. 33. 1786, nom. rejic.; *Tourrettia volubilis* J.F. Gmel., Syst. nat. 2: 940. 1791.

Description. Herbaceous vine; stems tetragonal. Leaves biternate, often with a fine, many-branched tendril; leaflets ovate to elliptic-ovate, to 8 cm long and 5.5 cm wide, glabrous; tip acuminate to acute; base obtuse to acute; margins serrate. Inflorescence spicate, short-pedicellate. Calyx red to scarlet, to 1.5 cm long, hirsute, 2-parted more than half their length; Corolla bilabiate, to 1.8 cm, purplish-greenish; tube glabrous internally, pubescent externally, to 1.2 cm long; lobes fused, forming hood over anthers but open below. Fruit a capsule, to 4.5 cm long, ovoid, echinate. Seeds small, to 7 mm long and 5 mm wide, winged.

Phenology. Flowers in March, September, October, and December. Fruits in January, March, June, September, October, and December.

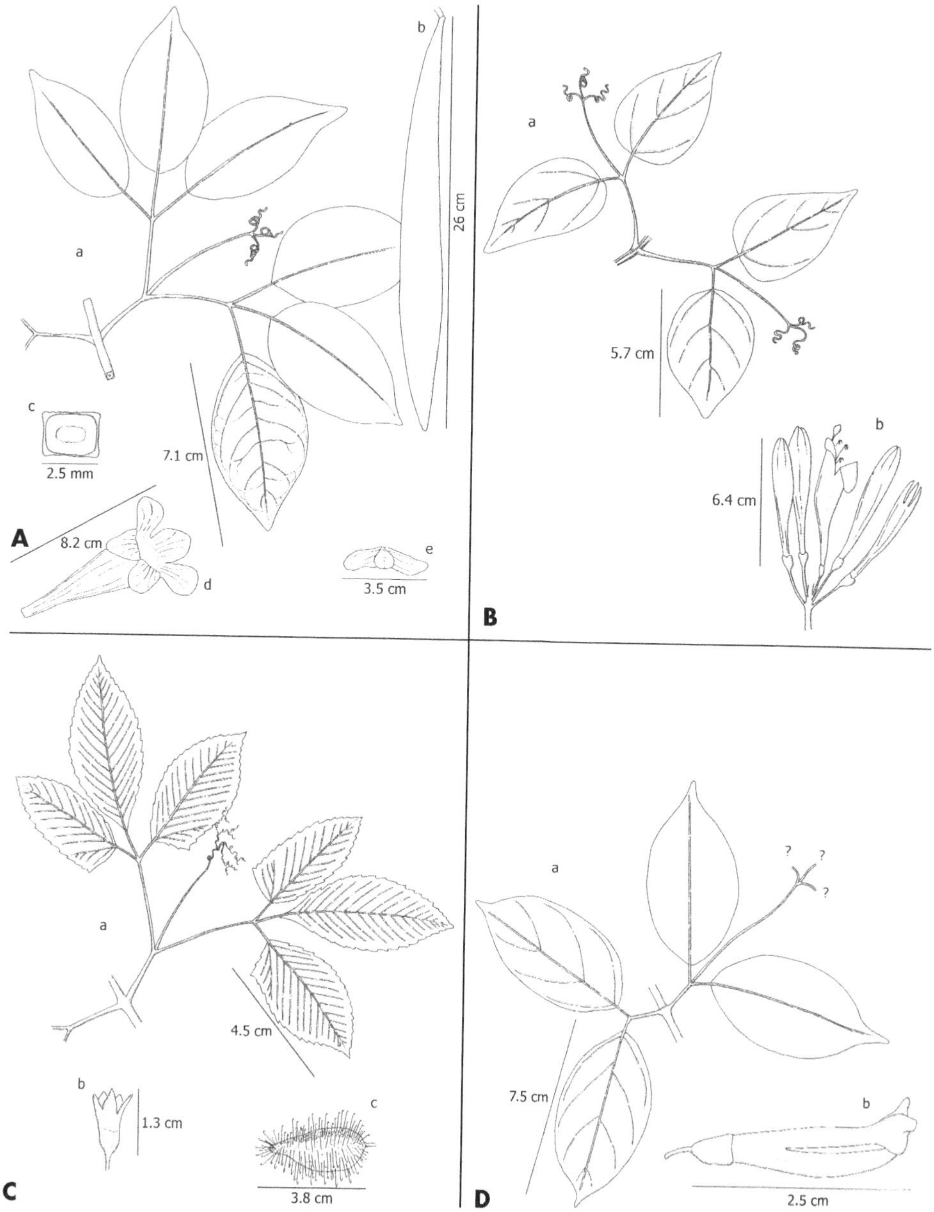

Fig. 3.6. A. *Pleonotoma variabilis:* **(a)** leaf and tendril, **(b)** fruit, **(c)** stem cross-section, **(d)** corolla, **(e)** seed (Leaf, stem cross-section, and corolla based on Foster 2181, DUKE; Fruit and seed based on Moreno 19464, MO). **B.** *Pyrostegia venusta:* **(a)** leaves and tendrils, **(b)** flowers. (Based on Khan et al. 900, CR). **C.** *Tourrettia lappacea:* **(a)** leaf and tendril, **(b)** flower, **(c)** fruit (Leaf and flower based on Lent 3247, DUKE; Fruit based on Hammel 19037, MO). **D.** *Tynanthus macranthus:* **(a)** leaves and tendril, **(b)** flower (Based on Lent 42, F, and Mori and Kallunki 6191, US).

Distribution. Southern Mexico to Argentina, generally from 600 to 2,500 m. In Costa Rica found from 1,000 to 2,000 m.

Notes. Sometimes known as *pega-pega* in El Salvador (Standley and Calderón 1925).

17. GENUS *TYNANTHUS*

Only one species of *Tynanthus* occurs in our area, the rare *T. macranthus*. It is recognized principally by the tannish pubescence on abaxial leaf surfaces and white corollas. The tendrils are probably uncinate, but the species is too poorly known to be certain. Our only other genera with uncinate-trifid tendrils, *Melloa* and *Macfadyena*, are glabrous on abaxial leaf surfaces (or puberulous only on the principal veins) and have yellow corollas.

1. ***Tynanthus macranthus*** L.O. Williams, Fieldiana, Bot. Ser. 31: 250. 1967. Figure 3.6D.

Description. Liana; stem cross-section unknown; branchlets subterete to terete, puberulous to pubescent; interpetiolar gland fields present; interpetiolar ridge present. Leaves bifoliolate; leaflets more or less ovate or elliptic, to 15 mm long and 6.2 cm wide, glabrous above, puberulous to velutinous below, plate-shaped glands scattered below near midrib; tip acute or acuminate; base rounded (sometimes oblique); margins entire. Tendrils trifid, uncinate (?). Inflorescence a few-flowered dichasia, axillary. Calyx narrowly campanulate, to 1 cm long and 0.5 cm wide, pubescent. Corolla white, to 3 cm long; lobes to about 0.8 cm long, slightly pubescent internally, pubescent externally. Fruit unknown.

Phenology. Flowers in April and July. Fruiting times unknown.

Distribution. Costa Rica and Panama. In Costa Rica very local from 100 to 800 m.

4
CUCURBITACEAE

With 19 genera and 42 species, the Cucurbitaceae is the largest family of tendrillate vines in our area. The family is easily identified because the tendrils are borne at right angles to the petiole. Although sometimes similar in leaf shape, members of the Vitaceae and Passifloraceae can be separated by tendril position (opposite the leaf in Vitaceae; truly axillary and not at right angles in Passifloraceae).

The Vitaceae and Passifloraceae are both characterized by superior ovaries, whereas the Cucurbitaceae has inferior ovaries that develop into berrylike fruits, called a pepo. The flowers of many genera in the Cucurbitaceae are characteristic and can greatly facilitate identification of the taxa with highly variable leaves (fig. 4.1).

Some species and several genera may also be easily recognized vegetatively. Petiole glands are found only in *Lagenaria siceraria*. Pinnatifidly or bipinnatifidly lobed leaves generally indicate *Citrullus lanatus* (*sandía* or watermelon), although the leaves of some *Cucumis* are also occasionally somewhat pinnatifidly lobed. Tendrils curling both above and below the bifurcation are found only in *Fevillea cordifolia* (a woody liana) and *Sicydium* (herbaceous vines).

The combination of simple leaves, conspicuous tufts of hair at the tip of the petiole, and patelliform glands below near the leaf base indicates *Tecunumania quetzalteca*. Other taxa with patelliform glands but lacking tufts of hair on the petiole are *Cionosicys*, *Cayaponia*, *Selysia*, and *Cyclanthera cogniauxii*. *Cyclanthera multifoliolata* also has glands on leaf undersurfaces and tufts of hair on the petiole (or lower laminar), but is distinguished from *Tecunumania* by compound leaves. In *Cionosicys* and *Cayaponia*, the glands are frequently crowded near the base on leaf undersurfaces, whereas in *Selysia* they are scattered throughout the lower surface. Our species of *Cyclanthera* with leaf glands are distinguished from other glandular taxa by a horizontal ring of anthers.

Our species of *Cyclanthera* (except *C. cogniauxii*, *C. dissecta*, and *C. explodens*) are characterized by 3–5-foliolate leaves with strongly dentate-serrate margins. Some species of *Cayaponia*, *Gurania*, and *Psiguria* also have trifoliolate leaves, but are distinguished by entire (or at least not dentate-serrate) margins or the presence of glands below near the leaf base.

No specimens of cultivated species of *Cucumis* and *Luffa* have been seen at elevations above 700 m in Costa Rica. However, because some *Cucumis* and *Luffa* have been collected above 900 m in neighboring Nicaragua and could potentially occur in our area, they are included in the generic key. *Cucurbita moschata* has been excluded, as it is known only in cultivation.

Many genera in this family are in need of further study, and our treatments are tentative. See Jeffrey (1978, 1990) for general classification and Lira and Chiang (1992) for *Sechium*. Synonymy follows Jeffrey (1971, 1978) and Nee (1993) when possible, Hampshire (1992a) for *Gurania*, and otherwise Wunderlin (1978). For further information see Dieterle (1976), Hammel and D'Arcy (1981), Kearns (1992), and Wunderlin (1976).

In addition to the keys, tables 4.1 and 4.2 provide help in identification by grouping various taxa by diagnostic characters. These tables do not include all Cucurbitaceae treated in the keys, and are intended solely as an aid to rapid identification of the more distinct taxa.

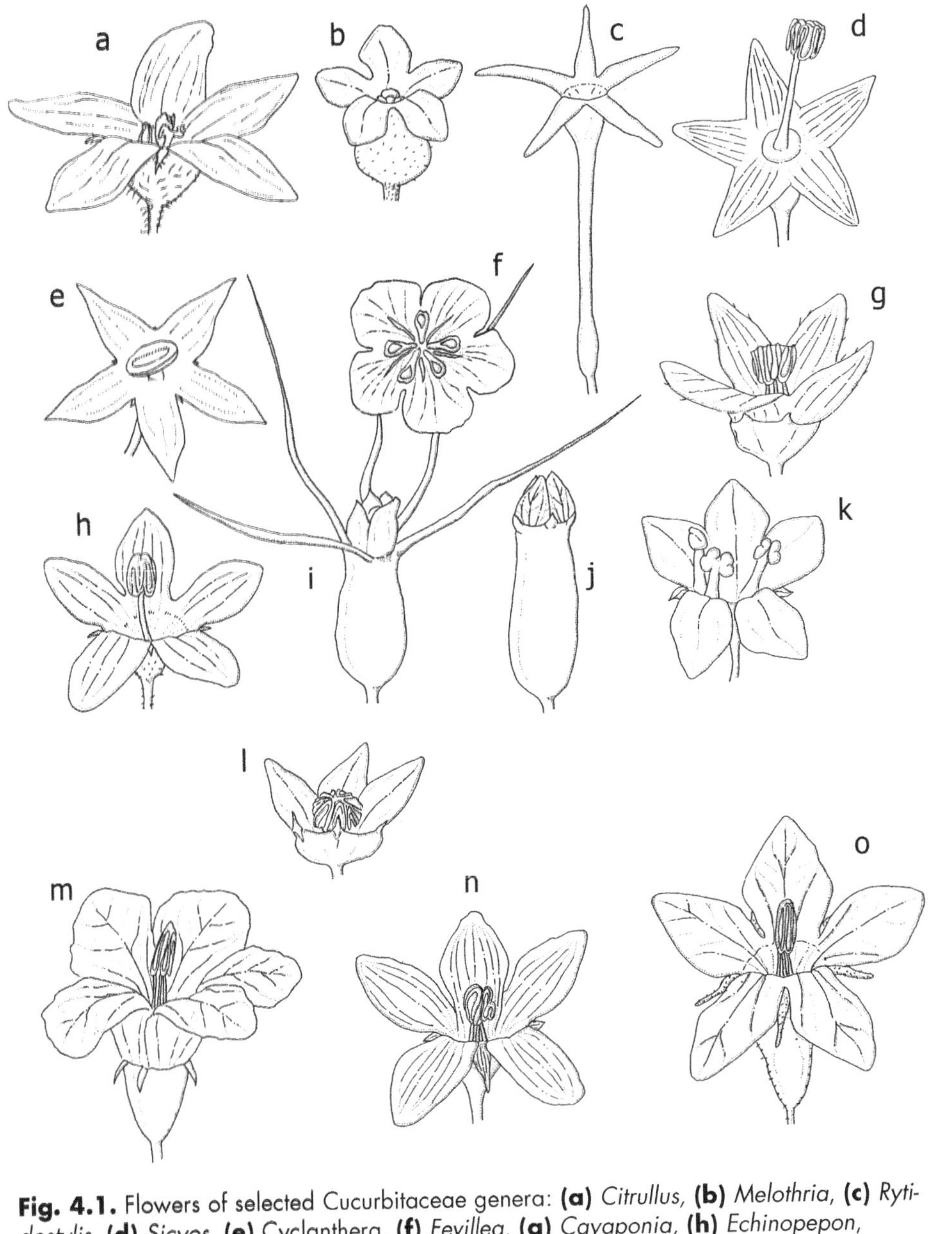

Fig. 4.1. Flowers of selected Cucurbitaceae genera: **(a)** *Citrullus*, **(b)** *Melothria*, **(c)** *Rytidostylis*, **(d)** *Sicyos*, **(e)** Cyclanthera, **(f)** *Fevillea*, **(g)** *Cayaponia*, **(h)** *Echinopepon*, **(i)** *Gurania*, **(j)** *Psiguria*, **(k)** *Sicydium*, **(l)** *Elateriopsis*, **(m)** *Lagenaria*, **(n)** *Momordica*, **(o)** *Polyclathra*. (Flowers not drawn to same scale.)

KEY TO CUCURBITACEAE USING PRIMARILY STAMINATE FLOWER AND VEGETATIVE CHARACTERS

1. Tendrils bifid, coiling both above and below the bifurcation
  2. Stamens 5; plant frequently a liana; ovary 3-locular... *Fevillea*, p. 55.
  2. Stamens 3; plant frequently an herbaceous vine; ovary 1-locular... *Sicydium*, p. 72.
1. Tendrils simple to many-branched, coiling only above the bifurcation
  3. Leaves pinnately lobed (infrequently in *Cucumis*)
    4. Tendrils 2–5-branched; staminate flowers solitary... *Citrullus*, p. 48.
    4. Tendrils simple; staminate flowers in fascicles or panicles... *Cucumis*, not treated.
  3. Leaves not pinnately lobed
    5. Petioles with conspicuous, raised glands near apex (sometimes found on extreme basal margin of leaf)... *Lagenaria*, p. 58.
    5. Petioles eglandular (if glands present near leaf base, then glands sessile, patelliform, and occurring on the leaf surface not the margin)
      6. Plant with combination of conspicuous tufts of hair near petiolar apex *and* scattered, patelliform glands below near leaf base... *Tecunumania*, p. 73.
      6. Plant lacking the above combination
        7. Leaves with patelliform glands below on leaf surface near base or scattered below
          8. Anthers in a horizontal ring... *Cyclanthera*, p. 49.
          8. Anthers not in a horizontal ring
            9. Anthers essentially straight or only slightly curved at end; glands scattered below throughout leaf surface... *Selysia*, p. 70.
            9. Anthers conduplicate or flexuous; glands below crowded near leaf base
              10. Fruit baccate, fibrous... *Cayaponia*, p. 43.
              10. Fruit peponiform, fleshy... *Cionosicys*, p. 48.
        7. Leaves lacking patelliform glands below on leaf surface near base
          11. Staminate inflorescence a solitary flower
            12. Corolla yellow; anthers conduplicate or flexuous... *Momordica*, p. 61.
            12. Corolla white; anthers straight or merely slightly curved at ends... *Polyclathra*, p. 62
          11. Staminate inflorescence with several flowers
            13. Corolla orange to red
              14. Calyx orange or reddish, lobes frequently at least 4 mm long (sometimes 2–3 times as long as the calyx tube)... *Gurania*, p. 55.
              14. Calyx green, lobes 1.5 mm long or less (always a fraction of the calyx tube length)... *Psiguria*, p. 62.
            13. Corolla white, greenish, to yellow
              15. Flowers with 10 nectaries at base of hypanthium... *Sechium*, p. 66.
              15. Flowers lacking nectaries at base of hypanthium
                16. Filaments distinct or connate only near base
                  17. Anthers straight or only slightly curved at ends... *Melothria*, p. 60.
                  17. Anthers conduplicate or flexuous
                    18. Petals connate, at least in lower part... *Cucumis*, not treated.

18. Petals free to base... *Luffa*, not treated.
16. Filaments connate
19. Anthers in a horizontal ring... *Cyclanthera*, p. 49.
19. Anthers vertically plicate or arched
20. Calyx elongate, narrowly cylindric... *Rytidostylis*, p. 66.
20. Calyx not elongate, more or less campanulate
21. Corolla salverform; lobes spreading... *Sicyos*, p. 72.
21. Corolla campanulate; lobes spreading or not
22. Stems and pedicels generally short hirsute or pilose to pilosulous, with conspicuous spreading hairs... *Echinopepon*, p. 54.
22. Stems and pedicels glabrous or puberulent with minute, generally appressed hairs... *Elateriopsis*, p. 54.

KEY TO CUCURBITACEAE USING PRIMARILY PISTILLATE FLOWER, FRUIT, AND VEGETATIVE CHARACTERS

1. Tendrils bifid, coiling both above and below the bifurcation
2. Ovary 3-locular; stigmas forming a 3-lobate, capitate head; fruit frequently greater than 10 cm wide; many-seeded, seeds more than 4 cm wide; plant frequently a liana... *Fevillea*, p. 55.
2. Ovary 1-locular; stigmas linear, spreading; fruit frequently less than 1.5 cm wide; 1-seeded, seeds generally less than 0.8 cm wide; plant frequently an herbaceous vine... *Sicydium*, p. 72.
1. Tendrils simple to many-branched, coiling only above the bifurcation
3. Leaves pinnately lobed (infrequently in *Cucumis*)
4. Tendrils 2–5-branched; ovary lacking tubercles or small, soft spines... *Citrullus*, p. 48.
4. Tendrils simple; ovary with tubercles or small, soft spines... *Cucumis*, not treated.
3. Leaves not pinnately lobed
5. Petioles with conspicuous, raised glands near apex (sometimes on extreme basal margin of leaf)... *Lagenaria*, p. 58.
5. Petioles eglandular or with sessile, patelliform glands on abaxial leaf surface, but not on margin
6. Plant with combination of simple leaves, conspicuous tufts of hair near petiole apex *and* patelliform glands scattered on abaxial surface near base... *Tecunumania*, p. 73.
6. Plant lacking the above combination
7. Leaves with glands on abaxial surface near base
8. Seeds triangular; glands scattered throughout abaxial leaf surface, not clustered only near base... *Selysia*, p. 70.
8. Seeds more or less ovate to elliptical; glands frequently clustered on abaxial leaf surface near base
9. Fruit setiferous to echinate, explosively or elastically dehiscent... *Cyclanthera*, p. 49.
9. Fruit smooth, indehiscent

10. Fruit baccate, fibrous... *Cayaponia*, p. 43.
10. Fruit peponiform, fleshy... *Cionosicys*, p. 48.
7. Leaves lacking glands on abaxial surface near base
11. Corolla orange or red
12. Calyx orange or reddish; lobes frequently at least 4 mm long (sometimes 2–3 times as long as calyx tube)... *Gurania*, p. 55.
12. Calyx green; lobes about 1.5 mm long or less (always shorter than calyx tube)... *Psiguria*, p. 62.
11. Corolla white, greenish, or yellow
13. Leaves 3–5-foliolate... *Cyclanthera*, p. 49.
13. Leaves simple (frequently lobate, but not compound)
14. Pistillate flowers or fruits in short-pedunculate clusters... *Sicyos*, p. 72.
14. Pistillate flowers or fruits not in short-pedunculate clusters
15. Flowers with 10 nectaries at base of hypanthium; ovule 1... *Sechium*, p. 66.
15. Flowers lacking nectaries at base of hypanthium; ovules more than 1
16. Ovary frequently tuberculate or with soft spines; fruit tuberculate or sparsely to densely echinate
17. Pedicels of inflorescences subtended by reniform, clasping bract; seeds red-arillate... *Momordica*, p. 61.
17. Pedicels of inflorescences not subtended by bract; seeds not red-arillate
18. Fruit explosively dehiscent
19. Pistillate flower flattened or shallowly cup-shaped... *Cyclanthera*, p. 49.
19. Pistillate flower tubular-elongate... *Rytidostylis*, p. 66.
18. Fruit indehiscent or if dehiscent, not explosive
20. Fruit lacking terminal beak, indehiscent, sparsely echinate to tuberculate... *Cucumis*, not treated.
20. Fruit with terminal beak, operculate at apex, densely echinate... *Echinopepon*, p. 54.
16. Ovary not tuberculate, lacking spines; fruit neither tuberculate nor echinate
21. Pedicels subtended by triangular, glanduliferous bract... *Luffa*, not treated.
21. Pedicels not subtended by triangular, glanduliferous bract
22. Staminate inflorescence a solitary flower; petals white; fruit indehiscent... *Polyclathra*, p. 62.
22. Staminate inflorescence with several flowers; petals greenish-white or white to yellow or greenish-yellow; fruit indehiscent or not
23. Tendrils bifid to many-branched; corolla greenish-white; fruit dehiscent... *Elateriopsis*, p. 54.

23. Tendrils simple; corolla yellow or white; fruit indehiscent
    24. Plant cultivated; corolla yellow; fruit globose, surface roughly netted, not smooth, 8–16 cm long... *Cucumis melo*, not treated.
    24. Plant spontaneous; corolla yellow or white; fruit ovoid or subglobose, surface essentially smooth, 7 cm long or less... *Melothria*, p. 60.

Table 4.1. Selected Cucurbitaceae Taxa Grouped by Reproductive Characters

Character	Taxa
Staminate flower with 5 stamens	*Echinopepon* *Fevillea* *Elateriopsis* *Sechium* *Cucurbita* (only cultivated)
Staminate inflorescence containing a single flower	*Cayaponia macrocalyx* (corolla greenish-yellow) *Cionosicys macranthus* (leaves with sessile, patelliform glands below near leaf base) *Citrullus lanatus* (leaves pinnately lobed) *Lagenaria siceraria* (glandular petioles or lower leaf margin with raised glands) *Momordica charantia* (corolla yellow; seeds red-arillate) *Polyclathra cucumerina* (corolla white) *Selysia prunifera* (leaves with sessile glands scattered below throughout; seeds arrowhead-shaped) *Tecunumania quetzalteca* (leaves with tufts of hair at petiolar apex and sessile, patelliform glands below crowded near leaf base)
Fruit one-seeded	*Sechium* *Sicydium* *Sicyos*
Fruit echinate	*Cucumis* *Cyclanthera* *Echinopepon* *Rytidostylis* *Sicyos*

Table 4.2. Selected Cucurbitaceae Taxa Grouped by Vegetative Characters

Character	Taxa
Petiole glandular or base of leaf margin with raised glands	*Lagenaria siceraria*
Leaf undersurfaces with sessile glands	*Cayaponia* (fruit baccate, fibrous) *Cionosicys macranthus* (fruit peponiform, fleshy) *Cyclanthera cogniauxii* and *C. multifoliolata* (anthers in a horizontal ring) *Selysia prunifera* (anthers straight or only slightly curved at ends) *Tecunumania quetzalteca* (petioles with tufts of hair near apex)
Leaves compound	*Cayaponia* (leaves with sessile, patelliform glands below near leaf base) *Cyclanthera* (margins strongly dentate-serrate) *Gurania* (calyx orange) *Psiguria* (calyx green)

1. GENUS *CAYAPONIA*

Like other genera in the Cucurbitaceae, *Cayaponia* needs revision. Species of *Cayaponia* differ largely in vegetative and floral characters. Fruit differences may also prove useful, but more study is needed. Five species are known from our area. The genus is one of five with patelliform glands on the abaxial leaf surface. Of these, *Tecunumania* is distinguished by tufts of hairs at the tip of the petiole. *Selysia* generally has glands scattered throughout the abaxial leaf surface and arrowhead-shaped seeds. *Cyclanthera* has anthers in a horizontal ring.

The closely related genus *Cionosicys* has traditionally been separated by its larger, fleshy, peponiform, many-seeded fruits, and petals connate over half their length. However, *Cayaponia macrocalyx* has solitary flowers of a size similar to *Cionosicys macranthus*. Other *Cayaponia* species (e.g., *C. glandulosa*) may produce fruits with up to 15 seeds. The degree of petal connation is also variable. Thus, the two genera are most reliably separated by the fruit type (peponiform and fleshy in *Cionosicys*, but baccate and fibrous in *Cayaponia*).

KEY TO *CAYAPONIA*

1. Mature leaves 3-foliolate (immature leaves may be simple)
    2. Leaves pubescent above (frequently scabrous), short tomentose below... *C. buraeavii*, p. 44.
    2. Leaves glabrous above and below... *C. granatensis*, p. 45.
1. Mature leaves simple, sometimes deeply lobate, but not 3-foliolate
    3. Leaves drying blackish; hypanthium about 3 cm wide; fruit frequently at least 4 cm wide... *C. macrocalyx*, p. 46.
    3. Leaves drying light to dark greenish; hypanthium to 1.7 cm wide; fruit to 1.7 cm wide
        4. Corolla of staminate flower frequently 1.5 cm long; fruit 3–15-seeded... *C. glandulosa*, p. 44.

4. Corolla of staminate flower frequently to 1 cm long; fruit 2–3-seeded... *C. racemosa*, p. 46.

1. *Cayaponia buraeavii* Cogn. in DC., Monogr. phan. 3: 791. 1881. Figure 4.2A.

Synonymy. *Cayaponia austin-smithii* Standl., Publ. Field Mus. Nat. Hist., Bot. Ser. 18: 1387. 1938.

Description. Vine; stems puberulent to glabrate. Leaves simple, to 13 cm long and 15 cm wide, unlobed or lobate, frequently with patelliform glands clustered near leaf base, scabrous to puberulous above, short-tomentose below; tip acute; base cordate. Tendrils simple or bifid, curling only above the bifurcation. Staminate flowers solitary, paired, or in racemes, axillary; calyx campanulate, to 8 mm long, somewhat villous internally, puberulous externally; corolla white, to 1 cm long; petals scattered villous internally, short-tomentose externally; stamens 3; filaments free; anthers connate. Pistillate flowers with calyx and corolla more or less as in staminate flowers; ovary subglobose, puberulous; stigmas 3 in a head. Fruit yellowish, to 2 cm long and 1.5 cm wide, ovoid, indehiscent. Seeds elliptic, to 9 mm long and 6 mm wide.

Phenology. Flowers from April through August. Fruits from August to November and in March.

Distribution. Costa Rica to Colombia and Venezuela. In Costa Rica found from 300 to 1,700 m.

2. *Cayaponia glandulosa* (Poepp. & Endl.) Cogn., in DC., Monogr. phan. 3: 755. 1881. Figure 4.2B.

Synonymy. *Bryonia glandulosa* Poepp. & Endl., Nov. Gen. sp. pl. 2: 56. 1838; *Trianosperma heterophylla* Naud., Ann. Sci. Nat., Bot. 4. 16: 144. 1862; *Cayaponia tomentosa* Cogn., Bull. Acad. Roy. Sci. Belgique, Sér. 2. 49: 194. 1880; *C. poeppigii* Cogn., Bull. Acad. Roy. Sci. Belgique, Sér. 2. 49: 195. 1880; *C. ottoniana* Cogn. in DC., Monogr. phan. 3: 755. 1881; *C. heterophylla* (Naud.) Cogn. in DC., Monogr. phan. 3: 758. 1881; *Arkezostis glandulosa* (Poepp. & Endl.) Kuntze, Revis. gen. pl. 1: 255. 1891; *A. ottoniana* (Cogn.) Kuntze, Revis. gen. pl. 1: 255. 1891; *A. heterophylla* (Naud.) Kuntze, Revis. gen. pl. 1: 255. 1891; *A. poeppigii* (Cogn.) Kuntze, Revis. gen. pl. 1: 255. 1891; *A. tomentosa* (Cogn.) Kuntze, Revis. gen. pl. 1: 255. 1891.

Description. Vine; stems pubescent to glabrate. Leaves simple, to 20 cm long and wide, unlobed or 3-lobate, frequently with patelliform glands clustered near leaf base, somewhat pubescent above and below (at least on veins); tip obtuse or acute; base cuneate to cordate. Tendrils simple or bifid, curling only above the bifurcation. Staminate flowers in racemes or panicles, axillary; calyx campanulate, to 1 cm long, pubescent externally; corolla greenish-white; corolla lobes oblong to lanceolate, to 1.5 cm long, tomentose internally, pubescent externally; stamens 3; filaments free; anthers connate in a head. Pistillate flowers solitary or in axillary racemes; calyx and corolla more or less as in staminate flowers; ovary ovoid, puberulent to glabrous; style to 1.5 cm long; stigma 3-parted. Fruit purplish-black, to 2 cm long and 1.5 cm wide, ellipsoid or subglobose, indehiscent. Seeds frequently 6 or more, to 6 mm long and 4 mm wide, obovate.

Phenology. Flowers in November. Fruits in April, July, and October.

Distribution. Costa Rica to Bolivia. In Costa Rica found from sea level to 1,100 m.

Notes. Without flowers or fruits easily confused with *C. racemosa*.

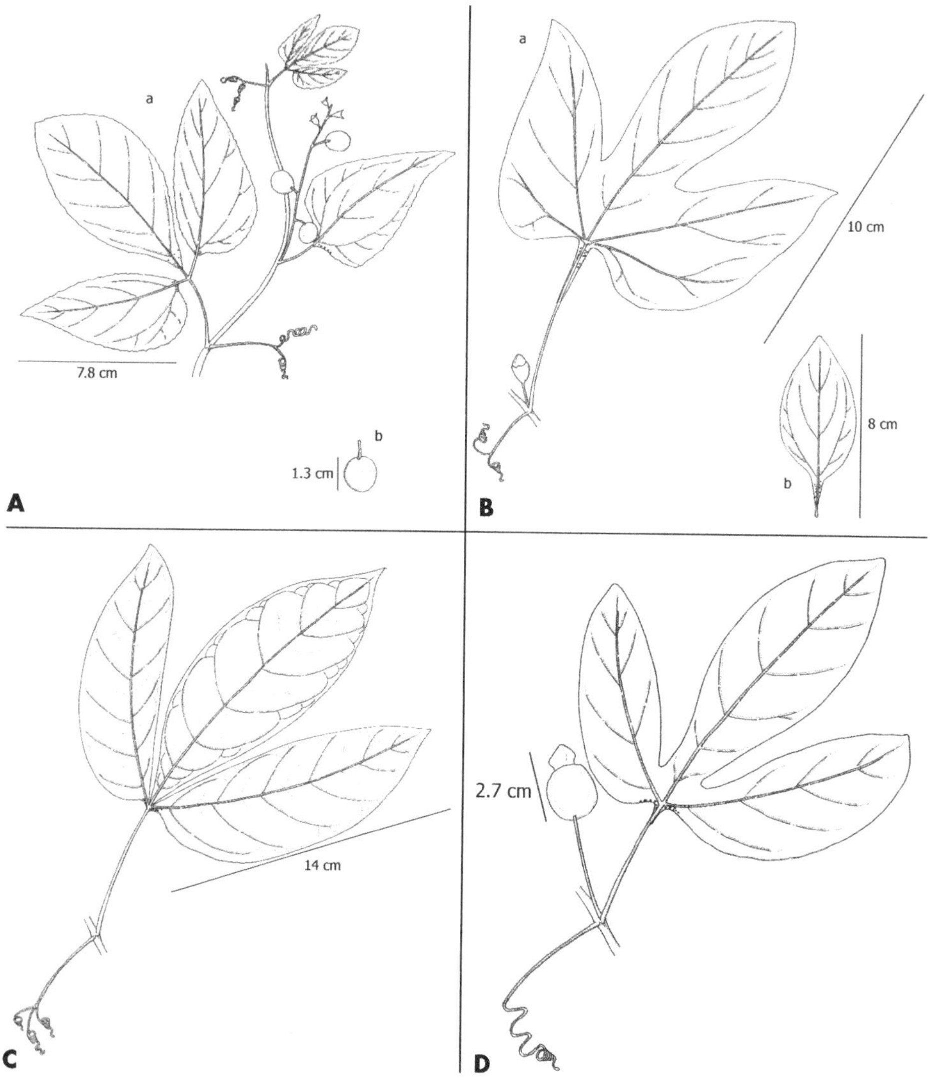

Fig. 4.2. A. *Cayaponia buraeavii:* **(a)** leaves, tendrils, and infructescence; **(b)** fruit (Based on Gómez 18795, DUKE). **B.** *Cayaponia glandulosa:* **(a)** leaf, tendril, and inflorescence; **(b)** leaf variation (Based on Pittier 4038, US). **C.** *Cayaponia granatensis:* leaf and tendril (Based on Foster 2208, US). **D.** *Cayaponia macrocalyx:* leaf, tendril, and inflorescence (Based on de Nevers et al. 4988, US).

3. *Cayaponia granatensis* Cogn., in DC., Monogr. phan. 3: 794. 1881. Figure 4.2C.

Synonymy. *Arkezostis granatensis* (Cogn.) Kuntze, Revis. gen. pl. 1: 255. 1891.

Description. Vine; stems glabrate. Leaves simple, to 22 cm long and wide, unlobed, 3-lobate or 3-foliolate, frequently with patelliform glands clustered near leaf base, glabrous above and below; tip acute to acuminate; base truncate to cordate. Tendrils trifid, curling only above the bifurcation. Staminate flowers frequently in racemes, axillary; calyx campanulate, to 2 mm long; corolla white; corolla lobes ovate or lanceolate, to 1.5 cm long, somewhat villous internally, pubescent externally; stamens 3; filaments free; anthers connate in a head. Pistillate flowers solitary or in axillary panicles; calyx and corolla more or less as in staminate flowers. Fruit reddish to orange, to 1.7 cm wide, globose, indehiscent. Seeds 2–3, to 1.5 cm long and 1.2 cm wide, ellipsoid.

Phenology. Flowers in January. Fruits from December through February.

Distribution. Costa Rica to Brazil and Peru. In Costa Rica found from 300 to 1,000 m.

4. *Cayaponia macrocalyx* Harms, Notizbl. Bot. Gart. Berlin-Dahlem 9: 994. 1926. Figure 4.2D.

Synonymy. *Cayaponia leucosticta* Standl., Publ. Field Mus. Nat. Hist., Bot. Ser. 18: 1388. 1939.

Description. Vine; stems glabrous. Leaves simple, to 13 cm long and 15 cm wide, 3-lobate, frequently with patelliform glands clustered near leaf base, glabrous above and below; tip acuminate; base cordate. Tendrils simple or bifid, curling only above the bifurcation. Staminate flowers frequently solitary, axillary; calyx urn-shaped to somewhat globose, about 3 cm wide, glabrous; calyx lobes to 3 mm long; corolla greenish-yellow; petals ovate or lanceolate, to 3 cm long, pubescent internally and externally; stamens 3; filaments free; anthers connate in a head; nectaries pouchlike near base of hypanthium. Pistillate flowers solitary, axillary; calyx and corolla more or less as in staminate flowers; ovary globose, to 2 cm long, glabrous; stigma 3-parted. Fruit green to purplish green, to 6 cm wide, globose, indehiscent. Seeds 3, to 1.7 cm wide, ovoid, basally tricornulate.

Phenology. Flowers in February and October. Fruits in December.

Distribution. Nicaragua to Ecuador. In Costa Rica found at about 1,100 m.

5. *Cayaponia racemosa* (Mill.) Cogn. in DC., Monogr. phan. 3: 768. 1881. Figure 4.3A.

Synonymy. *Bryonia racemosa* Mill., Gard. dict. 1768; *B. latebrosa* Aiton, Hortus kew. 3: 38. 1789; *Cayaponia latebrosa* (Aiton) Cogn. in DC., Monogr. phan. 3: 776. 1881; *Arkezostis latebrosa* (Aiton) Kuntze, Revis. gen. pl. 1: 255. 1891; *A. racemosa* (Mill.) Kuntze, Revis. gen. pl. 1: 255. 1891.

Description. Vine; stems glabrate. Leaves simple, to 13 cm long and 17 cm wide, unlobed or shallowly 3–5-lobate, frequently with patelliform glands clustered near leaf base, scabrous to hispidulous above, glabrous below, tip of lobes obtuse or acute; base cordate, blade sometimes decurrent on petiole. Tendrils simple or bifid, curling only above the bifurcation. Staminate flowers frequently in racemes, axillary; sepals more or less triangular, to 1 mm long; corolla greenish-white; petals villous externally, to 5 mm long; stamens 3; filaments free; anthers connate in a head. Pistillate flowers solitary or in axillary racemes; calyx and corolla more or less as in staminate flowers but smaller; ovary ovoid, glabrous; style to 6 mm long; stigma 3-parted. Fruit green, to 2 cm wide, ellipsoid, indehiscent. Seeds 2–3, to 6 mm wide, oblong or ovate.

Phenology. Flowers in August and October. Fruits in September, November, January, and February.

Distribution. Southern Florida, West Indies, Central America, and south to Colombia and Venezuela. In Costa Rica found from sea level to 1,600 m.

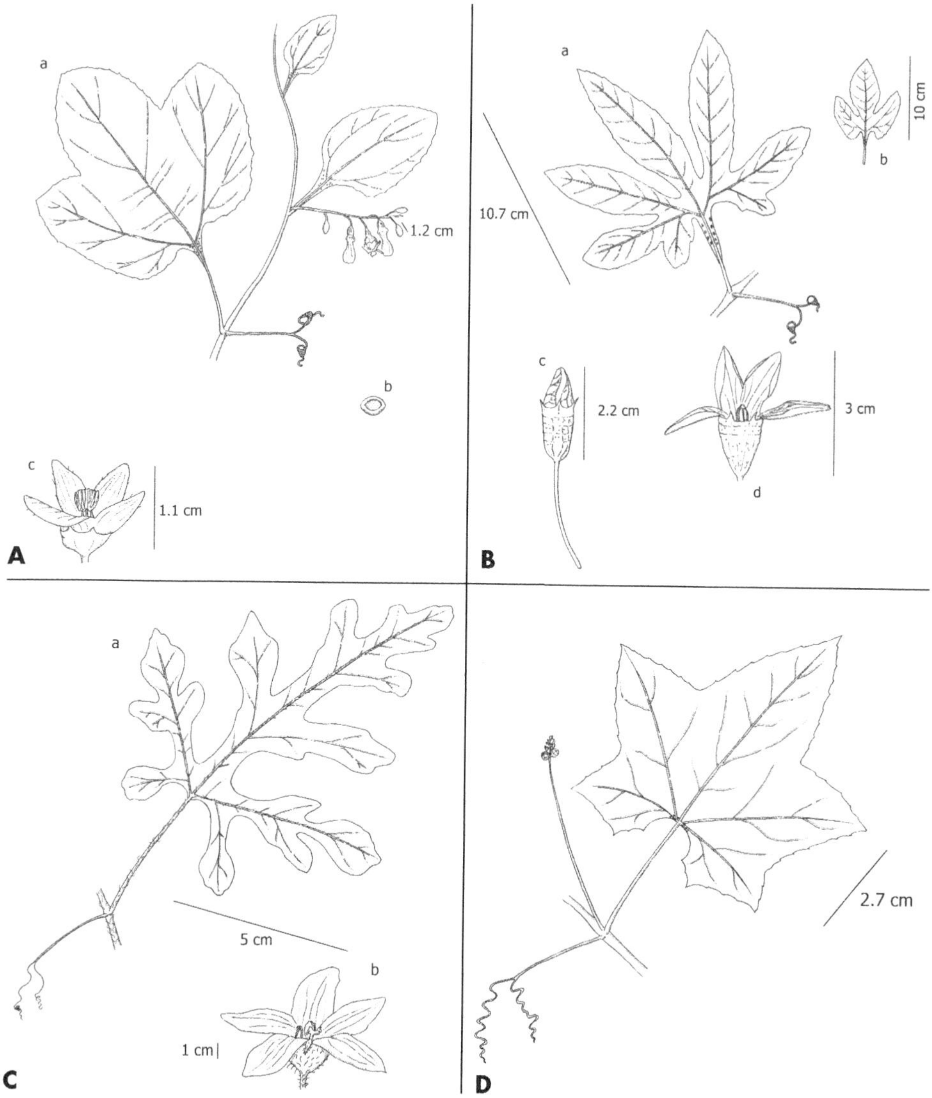

Fig. 4.3. A. *Cayaponia racemosa:* **(a)** leaves, tendril, and inflorescence; **(b)** close-up of laminar gland; **(c)** staminate flower (Leaves, tendril, inflorescence, and laminar gland based on Elias and Kirkbride, Jr. 1628, DUKE; Staminate flower based on Allen 1103, US). **B.** *Cionosicys macranthus:* **(a)** leaf and tendril, **(b)** leaf variation, **(c)** flower bud, **(d)** flower (Based on Krings 280, CR). **C.** *Citrullus lanatus:* **(a)** leaf and tendril, **(b)** staminate flower (Leaf and tendril based on Garnier A232, US; Staminate flower based on Ventura and López 1594, US). **D.** *Cyclanthera cogniauxii:* leaf, tendril, and inflorescence (Based on Lent 558, MO).

2. GENUS *CIONOSICYS*

Closely related to *Cayaponia*, *Cionosicys* is perhaps most easily distinguished by its peponiform, fleshy fruits (*Cayaponia* generally has baccate, fibrous fruits). In general, fruits are large and many-seeded in *Cionosicys*, but small and few-seeded in *Cayaponia*. Fruits of *Cayaponia macrocalyx* are exceptional and frequently reach larger sizes.

1. *Cionosicys macranthus* (Pittier) C. Jeffrey, Kew Bull. 25: 200. 1971. Figure 4.3B.

Synonymy. *Cayaponia macrantha* Pittier, Contr. U.S. Natl. Herb. 13: 118. 1910; *Cayaponia microdonta* S.F. Blake, Contr. U.S. Natl. Herb. 24: 27. 1922.

Description. Vine; stems glabrate. Leaves simple, to 13 cm long and 11 cm wide, 3–5-lobate, frequently scabrous to glabrous above and below, patelliform glands clustered near leaf base; base cordate, blade decurrent on petiole. Tendrils simple or bifid, curling only above the bifurcation. Staminate flowers typically solitary, axillary; calyx tubular-campanulate, to nearly 1.5 cm long; calyx lobes triangular, to 4 mm long; corolla white or cream (sometimes yellowish); stamens 3; filaments free; anthers connate. Pistillate flowers typically solitary, axillary; calyx and corolla more or less as in staminate flowers; ovary ovoid, puberulent; stigmas 3 in a peltate head. Fruit green, to 5 cm wide, frequently striped, globose, indehiscent. Seeds brown, to 1 cm wide, elliptic to ovate.

Phenology. Flowers in January, March, and October. Fruiting times unknown.

Distribution. Mexico to Panama. In Costa Rica found at about 1,100 m.

3. GENUS *CITRULLUS*

Sandía or watermelon (*Citrullus lanatus*) is generally found at lower elevations, although some plants have been found at 800 m. It is readily identified vegetatively by the pinnately or bipinnately lobed leaves. Some *Cucumis* species (also generally at lower elevations) may exhibit pinnate lobing, but never to the extent seen in *Citrullus*.

1. *Citrullus lanatus* (Thunb.) Matsum. & Nakai, Cat. Sem. Spor. Hort. Bot. Univ. Imp. Tokio 30, no. 854. 1916. Figure 4.3C.

Synonymy. *Cucurbita citrullus* L., Sp. pl. 1010. 1753; *Momordica lanata* Thunb., Prodr. fl. cap. 36. 1807; *Citrullus vulgaris* Schrad. in Eckl. & Zeyh., Enum. pl. afric. austral. 279. 1834.

Description. Vine; stems pubescent to glabrescent with age. Leaves pinnately or bipinnately lobed, to 15 cm long and 11 cm wide, frequently pilose above and below; tip acuminate or acute; base cordate. Tendrils frequently bifid, curling only above the bifurcation. Staminate flowers typically solitary, axillary; sepals oblong, to 4 mm long; corolla yellow, campanulate; corolla lobes oblong-ovate; stamens 3; filaments and anthers free. Pistillate flowers typically solitary, axillary; calyx and corolla as in staminate flowers; ovary ovoid, pubescent; ovules numerous, horizontal; styles 3, connate; stigmas 3, somewhat bilobate. Fruit large, green, globose to oblong, frequently with yellowish to reddish pulp, indehiscent. Seeds brownish, to 1 cm long and 0.6 cm wide.

Phenology. Flowers and fruits from November to January.

Distribution. Native to Africa, cultivated for its edible fruit. In Costa Rica found from sea level to 800 m.

4. GENUS *CYCLANTHERA*

Represented by seven species in our area, *Cyclanthera* is easily recognized by the anthers that are formed into a horizontal ring. Vegetatively the majority of species are 3–5-foliolate and characterized by strongly dentate- or crenate-serrate margins. Other genera with compound leaves generally have more or less entire or weakly serrulate margins (*Psiguria* and *Gurania*), or are characterized by sessile glands below near the leaf base (*Cayaponia*). Sessile glands in *Cyclanthera* are found only in *C. cogniauxii,* which has simple leaves, and *C. multifoliolata,* which has compound leaves. *C. explodens*, our only other common species with simple leaves, lacks glands altogether.

KEY TO *CYCLANTHERA*

1. Leaves simple
   2. Leaves with sessile glands below near leaf base... *C. cogniauxii*, p. 49.
   2. Leaves lacking sessile glands below near leaf base
      3. Staminate inflorescence with flowers only on upper two-thirds (or less) of main axis... *C. dissecta*, p. 50.
      3. Staminate inflorescence with flowers to base of main axis... *C. explodens*, p. 50.
1. Leaves 3–5-foliolate
   4. Leaves with numerous patelliform glands on abaxial surface near the base of leaflets... *C. multifoliolata*, p. 52.
   4. Leaves lacking glands
      5. Staminate inflorescence with flowers to base of main axis; androecial head sessile... *C. leptostachya*, p. 51.
      5. Staminate inflorescence with flowers only on upper two-thirds (or less) of main axis; androecial head stalked
         6. Plant cultivated; fruit indehiscent... *C. pedata*, p. 52.
         6. Plant spontaneous; fruit dehiscent
            7. Terminal leaflet pinnately lobed or peduncle less than 2 cm long... *C. dissecta*, p. 50.
            7. Terminal leaflet not pinnately lobed and peduncle greater than 3.5 cm long... *C. langaei*, p. 50.

1. ***Cyclanthera cogniauxii*** C.E. Jones & Kearns, Novon 4: 373. 1994. Figure 4.3D.

Description. Vine; stems glabrous to pubescent or tomentose (especially at nodes). Leaves typically 5–7-lobate, ovate to orbicular, glabrous to puberulent above on veins, glabrous below, frequently with small glands below near leaf base; tip acute to acuminate; base cordate; margins denticulate to dentate. Tendrils bifid. Staminate flowers in panicles or racemes, axillary; calyx to 5 mm wide; calyx lobes to 1 mm long; corolla greenish-white; corolla lobes ovate to lanceolate, to 3 mm long; anthers in a capitate, horizontal ring. Pistillate flowers solitary, axillary; calyx and corolla more or less as in staminate flowers; ovary ovoid, beaked, sometimes with scattered setiferous hairs; style 1; stigma subglobose. Fruit ovoid, to 2 cm long, setiferous to echinate. Seeds more or less ovoid, to 5 mm long, flattened.
Phenology. Flowers and fruits in January and May.
Distribution. Central Cordillera of Costa Rica from 1,500 to 2,500 m.

2. *Cyclanthera dissecta* (Torr. & A. Gray) Arn. & Hook., J. Bot. (Hooker) 3: 280. 1841. Figure 4.4A.

Synonymy. *Discanthera dissecta* Torr. & A. Gray, Fl. N. Amer. 1: 697. 1840; *Cyclanthera naudinia* Cogn. var. *tenuifolia* Cogn. in DC., Monogr. phan. 3: 381. 1881; *Cyclanthera tenuisepala* Cogn., Mém. Couronnés Acad. Roy. Sci. Belgique 28: 64. 1877; *Cyclanthera tonduzii* Cogn., Bull. Soc. Bot. Belg. 30: 274. 1891 or 1892(?).

Description. Vine; stems glabrous. Leaves typically 5-foliolate (sometimes 3-foliolate), to 12 cm long and wide; leaflets lanceolate, somewhat scabrous above, frequently glabrous below; tip acute; margin crenate to serrate. Tendrils frequently simple. Staminate flowers in panicles, axillary; calyx shallowly cup-shaped, to 5 mm wide; calyx lobes inconspicuous, subulate; corolla white to creamish; corolla lobes more or less lanceolate, less than 3 mm long, papillose internally, glabrous externally; anthers in a capitate, horizontal ring. Pistillate flowers solitary, axillary; calyx and corolla more or less as in staminate flowers, but smaller; ovary ovoid, beaked, setiferous; style 1; stigma subglobose. Fruit greenish, to 4 cm long, ovoid, setiferous, explosively dehiscent. Seeds ovoid, to 6 mm long, flattened.

Phenology. Flowers in April, October, and November.

Distribution. Southwestern United States to Panama. In Costa Rica found from 850 to 1,750 m.

3. *Cyclanthera explodens* Naud., Ann. Sci. Nat., Bot. Ser. 4. 12: 160. 1859. Figure 4.4B.

Synonymy. *Cyclanthera explodens* var. *intermedia* Cogn. ex Kuntze, Revis. gen. pl. 3: 103. 1898; *C. explodens* var. *trifida* Cogn. ex Kuntze, Revis. gen. pl. 3: 103. 1898.

Description. Vine; stems glabrous to puberulent. Leaves simple, to 10 cm long and wide, ovate to suborbicular, somewhat pustulate above and below, frequently puberulent along the veins; tip acute; base cordate; margins denticulate. Tendrils bifid, curling only above the bifurcation. Staminate flowers frequently in racemes, axillary; calyx shallowly cup-shaped, less than 2 mm wide; calyx lobes obsolete; corolla white; corolla lobes more or less lanceolate, papillose internally, glabrous externally; anthers in a capitate, horizontal ring. Pistillate flowers solitary, axillary; calyx and corolla more or less as in staminate flowers; ovary ovoid, beaked, sometimes scattered setiferous; style 1; stigma subglobose. Fruit greenish, to 4 cm long, dehiscent, ellipsoid, scattered setae, explosively dehiscent. Seeds ovoid, to 1 cm long, flattened.

Phenology. Flowering times unknown. Fruits from November through January.

Distribution. Mexico to Colombia. In Costa Rica found to 2,300 m.

4. *Cyclanthera langaei* Cogn., Mém. Couronnés Acad. Roy. Sci. Belgique 28: 67. 1877. Figure 4.4C.

Description. Vine; stems glabrous to puberulent, frequently woolly at the nodes. Leaves typically 3-foliolate, occasionally 5-foliolate; leaflets oblanceolate to elliptic or oblong but lower leaflets lobate, to 12 cm long and wide, puberulent above and below at least along midvein; tip acute or acuminate; margins crenate-serrulate to denticulate; petioles long villous at apex. Tendrils bifid. Staminate flowers in panicles, axillary; sepals obsolete; corolla green to greenish-yellow; corolla lobes more or less triangular; anthers in a capitate, horizontal ring. Pistillate flowers solitary, axillary; calyx and corolla more or less as in staminate flowers, but sometimes with larger corolla; ovary ovoid, beaked, scattered

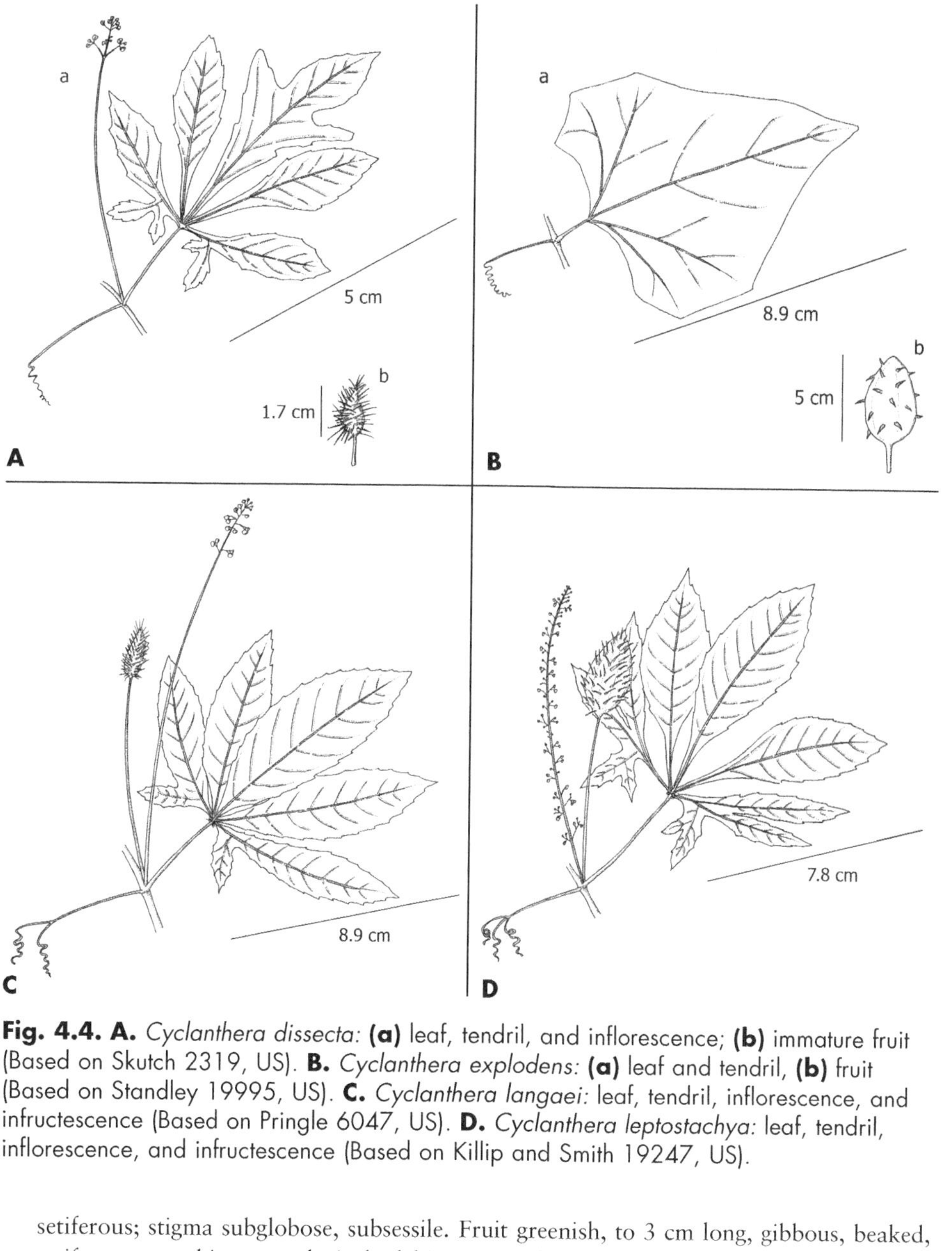

Fig. 4.4. A. *Cyclanthera dissecta:* **(a)** leaf, tendril, and inflorescence; **(b)** immature fruit (Based on Skutch 2319, US). **B.** *Cyclanthera explodens:* **(a)** leaf and tendril, **(b)** fruit (Based on Standley 19995, US). **C.** *Cyclanthera langaei:* leaf, tendril, inflorescence, and infructescence (Based on Pringle 6047, US). **D.** *Cyclanthera leptostachya:* leaf, tendril, inflorescence, and infructescence (Based on Killip and Smith 19247, US).

setiferous; stigma subglobose, subsessile. Fruit greenish, to 3 cm long, gibbous, beaked, setiferous, or echinate, explosively dehiscent. Seeds more or less ovoid, to 9 mm long.

Phenology. Flowers in April, July, August, November, and December. Fruits in February, July, and November.

Distribution. Mexico to Costa Rica. In Costa Rica found from 1,400 to 2,500 m.

5. *Cyclanthera leptostachya* Benth., Bot. voy. Sulphur 99. 1845. Figure 4.4D.

Description. Vine; stems puberulent to glabrate. Leaves typically 5-foliolate, occasionally 3-foliolate; leaflets more or less elliptic but lower leaflets lobate, to 15 cm long and 19 cm wide, somewhat scabrous to glabrate above and below; tip acute; margin serrate to dentate; petioles long villous at apex. Tendrils 3–5-fid. Staminate flowers in panicles, axillary; calyx shallowly cup-shaped, to 3 mm wide; calyx lobes minute; corolla greenish-white; corolla lobes more or less lanceolate, to 1 mm long, papillose internally, glabrous externally; anthers in a capitate, horizontal ring. Pistillate flowers solitary, axillary; calyx and corolla more or less as in staminate flowers, but smaller; ovary ovoid, beaked, setiferous; style 1; stigma subglobose. Fruit greenish, to 4.2 cm long, ovoid, setiferous, explosively dehiscent. Seeds ovoid, to 1.3 cm long, flattened.

Phenology. Flowers in January. Fruiting times unknown.

Distribution. Costa Rica to Peru. In Costa Rica found at about 750 m.

6. *Cyclanthera multifoliolata* Cogn., Mém. Couronnés Acad. Roy. Sci. Belgique 28: 66. 1877. Figure 4.5A.

Description. Vine; stems puberulent to glabrate. Leaves typically 3-foliolate, occasionally 5-foliolate; leaflets more or less elliptic but lower leaflets lobate, to 15 cm long and 20 cm wide, somewhat scabrous to glabrate above and below, numerous sessile glands below near base; tip acute; margin serrate to dentate; petioles villous to pubescent at apex. Tendrils 3–5-fid. Staminate flowers in panicles, axillary; calyx shallowly cup-shaped, to 3 mm wide; calyx lobes minute; corolla greenish-white; corolla lobes more or less lanceolate, to 1 mm long, papillose internally, glabrous externally; anthers in a capitate, horizontal ring. Pistillate flowers solitary, axillary; calyx and corolla more or less as in staminate flowers, but smaller; ovary ovoid, beaked, setiferous; style 1; stigma subglobose. Fruit greenish, to 4 cm long, ovoid, setiferous, explosively dehiscent. Seeds ovoid, to 1.3 cm long, flattened.

Phenology. Flowers in October and from January through March. Fruits in October, February, and March.

Distribution. Mexico to Peru. In Costa Rica found from 100 to 1,800 m.

7. *Cyclanthera pedata* (L.) Schrader, Index Sem. (Goettingen) 1831, 2. 1831. Figure 4.5B.

Synonymy. *Momordica pedata* L., Sp. pl. 2: 1009. 1753; *Anguria pedatisecta* Nees & Mart., Nova Acta Acad. Caes.-Leop. Carol. German. Nat. Cur. 12: 10. 1824; *A. pedisecta* Ser. in DC., Prodr. 3: 319. 1828, non Nees & Mart., 1824; *Momordica pedisecta* Ser. in DC., Prodr. 3: 319. 1828, pro syn; *Cyclanthera digitata* Arnott, J. Bot. (Hooker) 3: 280. 1841; *C. edulis* Naud., Belg. hort. 360. 1872; *C. pedata* var. *edulis* (Naud.) Cogn., in DC., Monogr. phan. 3: 826. 1881.

Description. Vine; stems glabrous. Leaves typically 5-foliolate (sometimes 3-foliolate), to 20 cm long and wide; lower leaflets lobate, lanceolate, glabrate above and below; tip acute; margins dentate-serrulate to serrate. Tendrils bifid. Staminate flowers in panicles, axillary; calyx shallowly cup-shaped, to 4 mm wide; calyx lobes inconspicuous or to 2 mm long; corolla white to creamish; corolla lobes more or less lanceolate, to 2 mm long, papillose internally, glabrous externally; anthers in a capitate, horizontal ring. Pistillate flowers solitary, axillary; calyx and corolla more or less as in staminate flowers; ovary ovoid, beaked, with or without scattered setae near the base; style 1; stigma subglobose. Fruit greenish, 5–16 cm long, ovoid, with or without sparse setae, indehiscent. Seeds ovoid, to 1.6 cm long, flattened.

Phenology. Flowers in January and August. Fruits in January and August.

Distribution. Mexico to Peru. In Costa Rica found about 1,300 m.

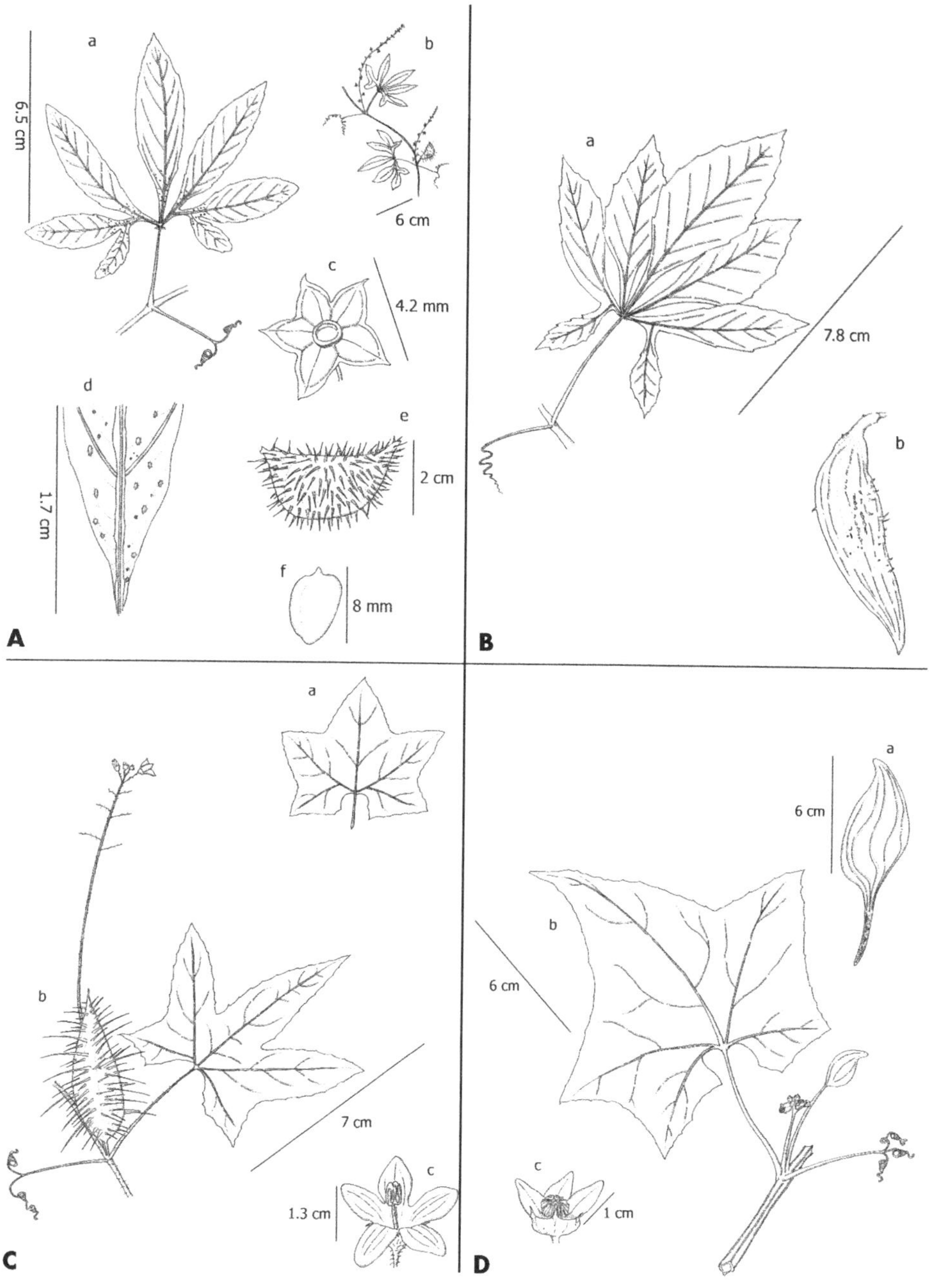

Fig. 4.5. A. *Cyclanthera multifoliolata:* **(a)** leaf and tendril, **(b)** habit, **(c)** staminate flower, **(d)** close-up of leaf base showing laminar glands, **(e)** fruit, **(f)** seed (Habit, leaf, leaf base, and tendril based on Tonduz 7036, US; Staminate flower based on McKee 11076, US; Fruit and seed based on Tonduz 9591, US). **B.** *Cyclanthera pedata:* **(a)** leaf and tendril, **(b)** fruit. (Leaf and tendril based on Cuatrecasas 19713, US. Fruit based on photo by Doyle and Pittier [Contr. U.S. Natl. Herb. 13 (4): 122. 1910]). **C.** *Echinopepon racemosus:* **(a)** leaf variation; **(b)** leaf, tendril, inflorescence, and infructescence; **(c)** staminate flower (Leaf, tendril, inflorescence, and infructescence based on Nee 23573, F; Leaf variation based on Breedlove and Raven 13157, F; Staminate flower based on Hinton et al. 8527, US). **D.** *Elateriopsis oerstedii:* **(a)** fruit; **(b)** leaf, tendril, inflorescence, and infructescence; **(c)** staminate flower with 2 petals removed (Leaf, tendril, inflorescence, and infructescence based on Krings 175, CR; Staminate flower based on Luteyn 12115, US).

Notes. Apparently native to the Andes, this species is commonly cultivated throughout the Neotropics for its edible fruit, and it sometimes escapes. In Costa Rica it is called *jaiba*.

5. GENUS *ECHINOPEPON*

Only one species of *Echinopepon* is known in our area, *Echinopepon racemosus*. It is characterized by a dehiscent (but not explosive), densely-echinate fruit.

1. ***Echinopepon racemosus*** (Steud.) C. Jeffrey, Kew Bull. 33: 357. 1978. Figure 4.5C.

Synonymy. *Momordica racemosa* Steud., Nomencl. bot. ed. 2: 155. 1851. Based on *Momordica muricata* Vell. Conc., Fl. flumin. 10, t. 94. 1835, non Willd.; *Echinopepon horridus* Naud., Ann. Sci. Nat., Bot. 5, 6: 19. 1866; *Echinocystis lanata* Cogn., Diagn. Cucurb. 2: 92. 1877.

Description. Vine; stems villous. Leaves simple, 3–5-lobate (infrequently 7-lobate), to 15 cm long and 10 cm wide, glabrate to sparsely pilose above, sparsely pilose below; tip acuminate to acute or obtuse; base cordate. Tendrils 2–3-fid, curling only above the bifurcation. Staminate flowers in racemes; sepals to 3 mm long; corolla white; petals ovate to somewhat triangular, more or less glandular internally; stamens 5; filaments connate; anthers partially or completely connate. Pistillate flowers solitary or paired, axillary; calyx and corolla as in staminate flowers; ovary ovoid, lanate, echinate; ovules 2–5, typically ascending, rarely horizontal; style to 3 mm long; stigmas globose. Fruit oblong to ovate, to 5 cm long and 2.5 cm wide (excluding beak and spines), densely echinate, beaked. Seeds more or less elliptical, to 8 mm long and 6 mm wide, verrucose.

Phenology. Flowering times unknown. Fruits in January.

Distribution. Mexico to Argentina. In Costa Rica found at about 1,350 m.

6. GENUS *ELATERIOPSIS*

Elateriopsis oerstedii is recognized by its glabrous to somewhat puberulous vegetative parts, whitish-green, campanulate flowers, stamen with fused filaments, vertically plicate or arched anthers, and a relatively smooth, explosively dehiscent fruit. This species lacks petiolar or laminar glands and petiolar tufts of hair.

1. ***Elateriopsis oerstedii*** (Cogn.) Pittier, Contr. U.S. Natl. Herb. 13: 125. 1910. Figure 4.5D.

Synonymy. *Cyclanthera oerstedii* Cogn. in DC., Monogr. phan. 3: 856. 1881; *Elateriopsis oerstedii* var. *biolleyi* Cogn. ex Pittier, Contr. U.S. Natl. Herb. 13: 126. 1910.

Description. Vine; stems glabrous to somewhat puberulous. Leaves simple, to 15 cm long and 9 cm wide, 3-lobate or angulate, glabrous to puberulous; tip acute to acuminate; base cordate. Tendrils 2–4-fid, curling only above the bifurcation. Staminate flowers borne in axillary racemes; calyx broadly campanulate; calyx lobes to 1 mm long; corolla greenish-white; corolla lobes ovate-lanceolate, to 1 cm long; stamens 5, connate. Pistillate flowers solitary, axillary; calyx and corolla as in staminate flowers; ovary obliquely fusiform, puberulent, 3-carpellate; styles 3, connate; stigmas forming a 3-lobate head. Fruit green, to 8 cm long and 4 cm wide, glabrous, elastically and longitudinally dehiscent. Seeds brownish, to 2.5 cm long and 2 cm wide, flattened; apex entire or notched.

Phenology. Flowers in September and from December through June. Fruits from September through June.

Distribution. Honduras (single collection), Costa Rica and Panama. In Costa Rica found from 100 to 1,520 m.

7. GENUS *FEVILLEA*

One of our few cucurbitaceous species that may reach the canopy, *Fevillea cordifolia* is unmistakable. Vegetatively it is recognized by its bifid tendrils that curl above and below the bifurcation, and glabrous, unlobed, or very shallowly lobed, cordate leaves. The fruits are greenish-yellow, marked by a circumferential scar near the apex, and frequently weigh up to 900 grams!

1. ***Fevillea cordifolia*** L., Sp. pl. 1013. 1753. Figure 4.6A.

Synonymy. *Trichosanthes punctata* L., Amoen. acad. 3: 423. 1756; *Fevillea scandens* L., Sp. pl. 1457. 1763; *F. trilobata* Reichard, Syst. Pl. 4: 253. 1780, non L., 1753; *F. punctata* (L.) Poir., in Lam., Encycl. 4: 418. 1796; *F. hederacea* Poir., in Lam., Encycl. 4: 419. 1796; *F. javilla* Kunth in H.B.K., Nov. Gen. sp. pl. 2: 124. 1817; *Nhandiroba scandens* (L.) Desc., Fl. méd. Antilles 8: 66. 1829; *Fevillea cordifolia* var. *hederacea* (Poir.) Cogn., in DC., Monogr. phan. 3: 943. 1881; *Nhandiroba cordifolia* (L.) Kuntze, Revis gen. pl. 1: 257. 1891.

Description. Vine or liana; stems glabrous. Leaves simple, to 16 cm long and 13 cm wide, more or less cordate, sometimes very shallowly lobate, glabrous above and below; tip acute or acuminate; base cordate. Tendrils bifid, curling above and below the bifurcation. Staminate flowers borne in panicles; sepals ovate, to 2 mm long; corolla yellow, to 4 mm long, each petal with a median uncinate appendage; stamens 5, free. Pistillate flowers with calyx and corolla as in staminate flowers; ovary globose, puberulent, 3-carpellate; styles 3, connate for at least part of their length; stigmas forming a capitate head. Fruit green to greenish-yellow, to 13 cm wide, globose, indehiscent, circumferential scar near apex. Seeds at least six per carpel, to 6 cm long and 5 cm wide.

Phenology. Flowers from December through March. Fruits in February and from May through July.

Distribution. West Indies, Central America and south to Peru. In Costa Rica found from sea level to 1,450 m.

8. GENUS *GURANIA*

The genus *Gurania* is closely related to *Psiguria* and is primarily separated by the color of the calyx (orange in *Gurania* and green in *Psiguria*). In our species, the calyx lobes are also always significantly longer than in *Psiguria* (at least 4 mm long in *Gurania* and 1.5 mm or less in *Psiguria*). Members of both genera have simple tendrils.

Four species of *Gurania* are known in our area, *G. coccinea*, *G. eriantha*, *G. makoyana*, and *G. tubulosa*, but *G. eriantha* has been tentatively included based upon Hampshire (1992a). Long thought to be dioecious, Condon and Gilbert (1988) discovered that plants in both genera are actually monoecious, but produce the male and female flowers at different times.

KEY TO *GURANIA*

1. Pedicels of staminate flowers 1–3 cm long; calyx tube of staminate flowers longer than 1 cm... *G. tubulosa*, p. 58.
1. Pedicels of staminate flowers less than 1 cm long; calyx tube of staminate flowers shorter than 1 cm
   2. Stems villous with hairs generally longer than 4 mm; staminate flowers subsessile... *G. eriantha*, p. 56.
   2. Stems glabrous to pubescent with hairs shorter than 4 mm; if villous then staminate flowers on pedicels at least 6 mm long
      3. Leaves simple or compound; lowest staminate flowers on pedicels 4 mm or more long; calyx lobes generally shorter than calyx tube or to 1.5 times as long (infrequently twice as long)... *G. coccinea*, p. 56.
      3. Leaves simple; lowest staminate flowers on pedicels 1–3 mm long; calyx lobes frequently 2–3.5 times as long as calyx tube... *G. makoyana*, p. 58.

1. *Gurania coccinea* Cogn., Mém. Couronnés Acad. Roy. Sci. Belgique 27: 42. 1876. Figure 4.6B.

Synonymy. *Gurania costaricensis* Cogn., Mém. Couronnés Mém. Acad. Roy. Sci. Belgique 27: 42. 1876; *Gurania racemifera* Standl., Publ. Field Mus. Nat. Hist., Bot. Ser. 18: 1398. 1938.

Description. Vine; stems pilose to glabrate. Leaves simple and unlobed to 3-lobate or 3-foliolate, to 18 cm long, pilose above and below; tip acute or acuminate; base cuneate. Tendrils simple. Staminate flowers in umbeloid racemes, axillary; pedicels 0.4–1.5 cm long; calyx orange, tubular, puberulent or papillose internally, pilose externally; corolla yellow, papillose internally and externally; anthers 2, to 3 mm long, free, replicate at base. Pistillate flowers with calyx and corolla more or less as in staminate flowers; ovary fusiform; styles 2, connate. Fruit green, to 6 cm long, spotted, oblong to fusiform. Seeds white, to 1 cm long and 0.5 cm wide, ovoid.

Phenology. Flowers and fruits throughout the year.

Distribution. Nicaragua to Panama. In Costa Rica found from 10 to 1,500 m.

2. *Gurania eriantha* (Poepp. & Endl.) Cogn., Mém. Couronnés Acad. Roy. Sci. Belgique 27: 16. 1876. Figure 4.6C.

Synonymy. *Anguria eriantha* Poepp. & Endl., Nov. Gen. sp. pl. 2: 52. 1838; *Gurania hirsuta* Cogn., Bot. Jahrb. Syst. 23: 134. 1896.

Description. Vine; stems pilose to glabrate. Leaves simple, to 22 cm long, unlobed to 3-lobate to 3-foliolate, pilose above and below; tip acute or acuminate; base cordate. Tendrils simple. Staminate flowers in umbeloid racemes, axillary, sessile or subsessile; calyx orange, urn-shaped, glabrous internally, pilose externally; corolla yellow, papillose internally and externally; anthers replicate at ends, to 3 mm long. Pistillate flowers with calyx and corolla more or less as in staminate flowers; ovary ellipsoid; styles 2, connate. Fruit green to brown, to 5 cm long, oblong. Seeds brownish, to 1.2 cm long and 0.6 cm wide, ovoid, flattened.

Phenology. Flowering and fruiting times unknown.

Distribution. Costa Rica and Panama.

Notes. This species is tentatively included based on the concept of Hampshire (1992a). However, the only specimen seen (a photograph of Rothschub 8996, Berlin), did not seem

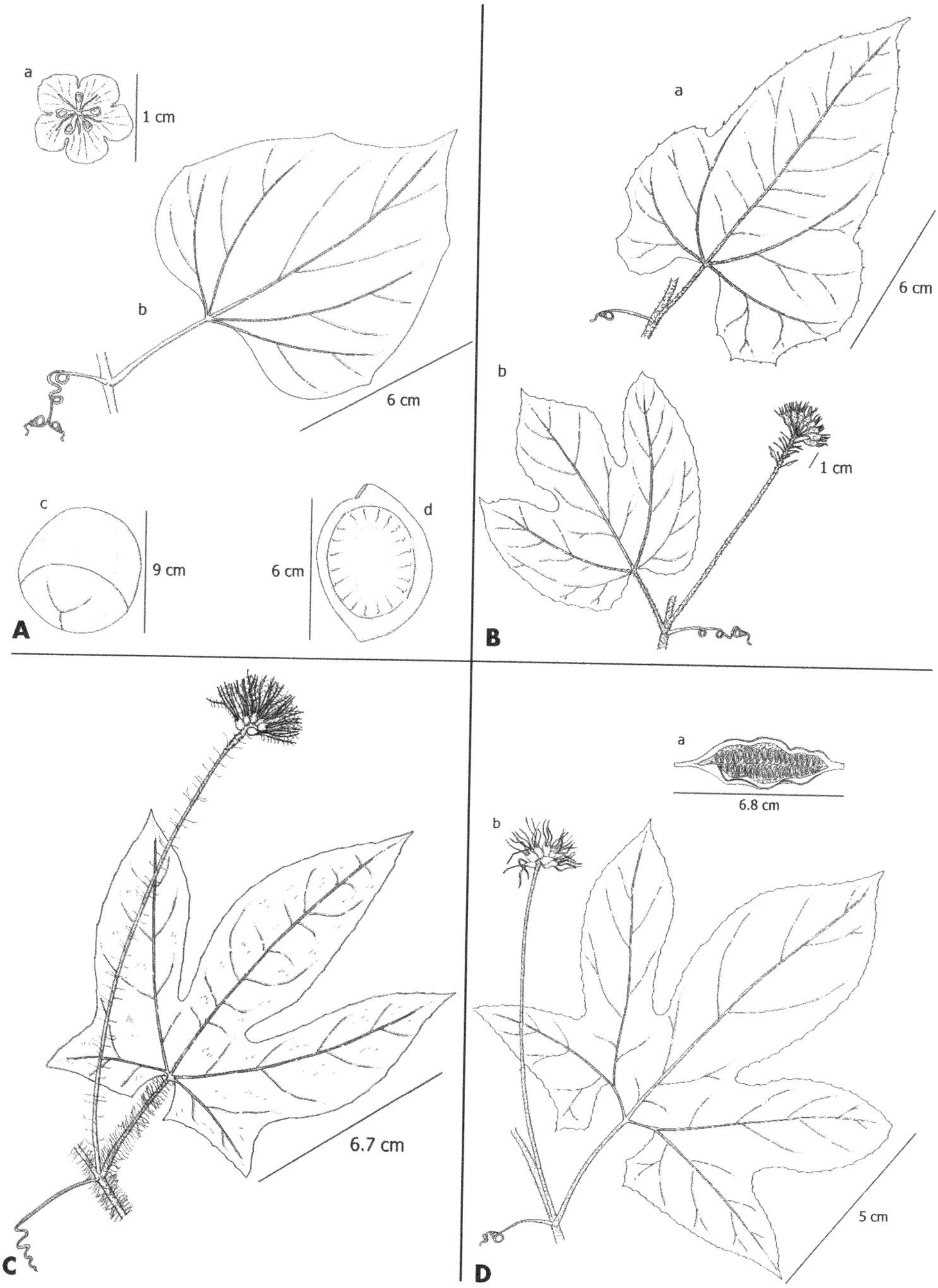

Fig. 4.6. A. *Fevillea cordifolia:* **(a)** flower, **(b)** leaf and tendril, **(c)** fruit, **(d)** seed (Leaf and tendril based on Herréra 2434, MO; Flower based on Busey 440, US; Fruit and seed based on Krings 28, CR). **B.** *Gurania coccinea:* **(a)** leaf and tendril; **(b)** leaf, tendril, and inflorescence (Based on Carvajal U. 442, DUKE). **C.** *Gurania eriantha:* leaf, tendril, and inflorescence (Based on McDaniel 17984, F). **D.** *Gurania makoyana:* **(a)** opened fruit; **(b)** leaf, tendril, and inflorescence (Based on Taylor 3556, DUKE).

to fit her description. The plant is villous and simple-leaved, but with floral pedicels between 5 and 6 mm long, not subsessile. The long calyx lobes (to 2.5 cm) are reminiscent of *G. makoyana*, although the longer pedicels seem closer to *G. coccinea*.

3. *Gurania makoyana* (Lem.) Cogn., Mém. Couronnés Acad. Roy. Sci. Belgique 27: 17. 1876. Figure 4.6D.

Synonymy. *Anguria makoyana* Lem., Fl. Serres 3: tab. 222. 1847; *Gurania seemaniana* Cogn., Mém. Couronnés Acad. Roy. Sci. 9 Belgique 27: 35. 1876; *G. donnell-smithii* Cogn. ex Donn. Sm., Bot. Gaz. 16: 10. 1891; *G. levyana* Cogn., Diagn. Cucurb. 1: 26. 1876; *G. tonduziana* J.D. Sm., Bot. Gaz. 31: 112. 1901.

Description. Vine; stems glabrous to villous. Leaves simple, to 30 cm long, unlobed to 3–7-lobate, pilose above and below; tip acuminate; base cordate. Tendrils simple. Staminate flowers in umbeloid racemes, axillary; pedicels 1–3 mm long; calyx orange, urn-shaped, puberulent or papillose internally, pilose externally; corolla yellow, papillose internally and externally; anthers 2, to 6 mm long, free, replicate at base. Pistillate flowers with calyx and corolla more or less as in staminate flowers, but glabrous internally; ovary fusiform; styles 2, connate. Fruit green, to 6 cm long, spotted, oblong to fusiform. Seeds white, to 1 cm long and 0.5 cm wide, ovoid, flattened.

Phenology. Flowers in August, September, and from November through April. Fruits in November and January.

Distribution. Mexico to Panama. In Costa Rica found from 10 to 1,500 m.

4. *Gurania tubulosa* Cogn., Diagn. Cucurb. 1: 30. 1876. Figure 4.7A.

Synonymy. *Gurania megistantha* J.D. Sm., Bot. Gaz. 33: 251. 1902; *Gurania suberosa* Standl., J. Wash. Acad. Sci. 15: 461. 1925.

Description. Vine or liana, stems pilose. Leaves simple, to 22 cm long, unlobed or 3-lobate, pilose above and below especially along the veins; tip acute or acuminate; base cordate. Tendrils simple. Staminate flowers in umbeloid racemes, axillary; pedicels 1–3 cm long; calyx orange, tubular, puberulent to glabrate externally, puberulent or papillose internally; corolla yellow, papillose on both surfaces; anthers 2, free, replicate at base. Pistillate flowers with calyx and corolla more or less as in staminate flowers; ovary oblong; styles 2, connate. Fruit green, to 5 cm long, spotted, oblong. Seeds white, to 1 cm long and 0.5 cm wide, more or less ovoid.

Phenology. Flowers in January and May. Fruiting times unknown.

Distribution. Costa Rica and Panama. In Costa Rica found from 800 to 1,600 m.

9. GENUS *LAGENARIA*

The only species of *Lagenaria* in our area, *L. siceraria*, is easily recognized by the raised glands found near the apex of the petiole or on the extreme basal margin of the leaves. All other Cucurbitaceae species from our area have eglandular petioles. Five additional genera have sessile, patelliform glands, but these are typically found on leaf undersurfaces near the base, not on the margins (table 4.1).

1. *Lagenaria siceraria* (Molina) Standl., Publ. Field Mus. Nat. Hist., Bot. Ser. 3: 435. 1930. Figure 4.7B.

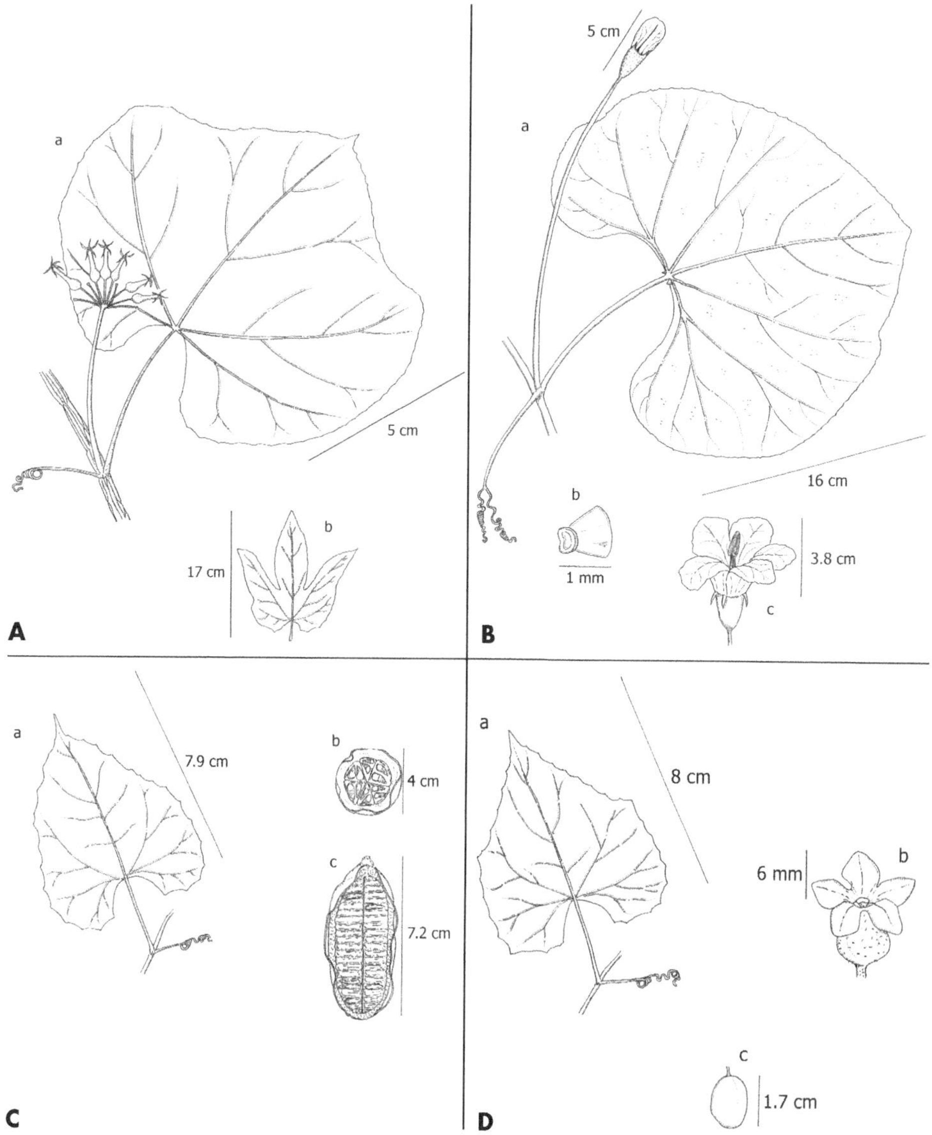

Fig. 4.7. A. *Gurania tubulosa:* **(a)** leaf, tendril, and inflorescence; **(b)** leaf variation (Based on Foster 654, DUKE). **B.** *Lagenaria siceraria:* **(a)** leaf, tendril, and inflorescence; **(b)** close-up of petiolar gland; **(c)** staminate flower (Leaf, tendril, inflorescence, and gland based on Lewis 1959, US; Staminate flower based on Pittier 5028, US). **C.** *Melothria dulcis:* **(a)** leaf and tendril, **(b)** cross-section of fruit, **(c)** longitudinal section of fruit (Based on Hammel 8346, DUKE). **D.** *Melothria pendula:* **(a)** leaf and tendril, **(b)** flower, **(c)** fruit (Based on Taylor 960, DUKE).

Synonymy. *Cucurbita lagenaria* L., Sp. pl. 1010. 1753; *Cucurbita siceraria* Molina, Sag. stor. nat. Chili 133. 1782.

Description. Vine; stems frequently pubescent. Leaves simple, to 15 cm long and 20 cm wide, broadly ovate to somewhat reniform, frequently somewhat angulate or shallowly lobate, frequently pubescent above and below along the veins; tip obtuse; base cordate; petiole with conspicuous glands near apex. Tendrils bifid, curling only above the bifurcation. Staminate flowers solitary, axillary; sepals more or less triangular (sometimes linear), to 4 mm long; corolla white, opening at night, 5-lobate; corolla lobes to 4 cm long; stamens 3, free. Pistillate flowers with calyx and corolla as in staminate flowers; ovary ovoid to cylindric, villous; ovules numerous, horizontal; stigmas 3. Fruit usually large, variable in size and shape. Seeds obovate to oblong, to 1.5 cm long, flattened, more or less bicornute at the ends; margins or submargins raised.

Phenology. Flowers and fruits in November.

Distribution. Cultivated and naturalized throughout tropical America. In Costa Rica found at about 1,100 m.

Notes. Known in Mexico as *cuchara* or *tecomate* and in Costa Rica as *calabaza*. The large, hard-shelled fruits have been used for a variety of utilitarian purposes in Central America including canteens.

10. GENUS *MELOTHRIA*

Three species of *Melothria* are found in our area, *M. dulcis*, *M. pendula*, and *M. scabra*. Vegetatively they are often extremely similar and cannot be reliably separated. All three exhibit simple tendrils, frequently pustulate leaf surfaces, eglandular petioles, yellow or white corollas, staminate inflorescences bearing several flowers, free filaments, and more or less straight anthers. In addition, leaf undersides lack patelliform glands near the base.

Among our species only *M. dulcis* has a white corolla (the remaining species have yellow corollas). In fruit, *M. pendula* may be recognized by its small, uniformly colored (not striped) fruits. Both other species have larger fruits (to 7 cm long), but those of *M. dulcis* have orange pulp, whereas *M. scabra* has whitish or greenish pulp.

KEY TO *MELOTHRIA*

1. Corolla white; fruit yellow or orange when ripe; pulp orange... *M. dulcis*, p. 60.
1. Corolla yellow; fruit green to black when ripe, sometimes striped; pulp not orange
   2. Fruit evenly green to black when ripe, to 2 cm long... *M. pendula*, p. 61.
   2. Fruit light and dark green striped when ripe, at least 2.5 cm long... *M. scabra*, p. 61.

1. ***Melothria dulcis*** Wunderlin, Phytologia 38: 220. 1978. Figure 4.7C.

Description. Vine; stems scabrous to glabrate. Leaves simple, to 14 cm long and 10 cm wide, pustulate and scabrous to glabrate above, strigose below mainly on the veins; tip acuminate; base cordate; margin undulate, denticulate. Tendrils simple. Staminate flowers in racemes, axillary; calyx narrowly campanulate, to 5 mm long, glabrous; calyx lobes more or less triangular, to 1 mm long; corolla white, to 3 mm long; corolla lobes oblong, pubescent internally and externally; stamens 3, inserted on hypanthium wall; anthers oblong, papillose at apex. Pistillate flowers solitary, axillary; calyx and corolla more or less as in staminate flowers; ovary 3-carpellate; ovules numerous, horizontal; styles 3, connate; stigmas 3, linear. Fruit yellow to orange with orange pulp, to 7 cm long and 6 cm wide, ovoid to subspherical, glabrous. Seeds ovoid, to 1 cm long.

Phenology. Flowering times unknown. Fruits from February through April, in July, and from September through October.

Distribution. Nicaragua to Panama. In Costa Rica found from sea level to 800 m.

Notes. The fruits are edible and very delicious.

2. *Melothria pendula* L., Sp. pl. 35. 1753. Figure 4.7D.

Synonymy. *Bryonia guadalupensis* Spreng., Syst. veg. 3: 580. 1881; *Melothria fluminensis* Gardner, J. Bot. (Hooker) 1: 173. 1842; *M. guadalupensis* (Spreng.) Cogn., in DC., Monogr. phan. 3: 580. 1881.

Description. Vine; stems scabrous to glabrate. Leaves simple, to 10 cm long and wide, cordate to sagittate (sometimes 3–5-lobate), pustulate and scabrous above and below; tip acuminate; base cordate; margins undulate, denticulate. Tendrils simple. Staminate flowers in racemes, axillary; calyx campanulate, glabrate; calyx lobes subulate, nearly 1 mm long; corolla yellow; corolla lobes oblong, papillose or pubescent internally and externally; stamens 3, inserted on hypanthium wall; anthers oblong. Pistillate flowers solitary, axillary; calyx and corolla more or less as in staminate flowers; ovary ellipsoid, glabrous, 3-carpellate; ovules numerous, horizontal; styles 3, connate; stigmas 3, linear. Fruit green to black, to 1.5 cm (rarely 2 cm) long and 1 cm wide, ovoid, glabrous. Seeds ovoid, to 5 mm long, white-sericeous.

Phenology. Flowers and fruits throughout the year.

Distribution. West Indies, Southern United States to Argentina. In Costa Rica found from sea level to 1,500 m.

3. *Melothria scabra* Naud., Ann. Sci. Nat., Bot. 6: 10. 1867. Figure 4.8A.

Synonymy. *Melothria donnell-smithii* Cogn., Bot. Gaz. 16: 9. 1891; *M. donnell-smithii* var. *hirtella* Cogn., Bot. Gaz. 16: 9. 1891; *M. donnell-smithii* var. *rotundifolia* Cogn., Bot. Gaz. 16: 9. 1891; *M. pendula* Mottet, Rev. hort. 660. 1900, non L., 1753.

Description. Vine; stems scabrous or pilose to glabrate. Leaves simple, to 10 cm long and wide, cordate, sometimes 3–5-lobate, pustulate, scabrous, and pilose above (at least on veins), pilosulous below mainly on veins; tip acuminate; base cordate; margin undulate, denticulate to dentate. Tendrils simple. Staminate flowers in racemes, axillary; calyx campanulate, pilosulous to glabrate; calyx lobes more or less triangular, to 1 mm long; corolla yellow; corolla lobes ovate or oblong, to 3 mm long, glabrate internally, pubescent externally; stamens 3, inserted on hypanthium wall; anthers subtetragonal. Pistillate flowers solitary, axillary; calyx and corolla more or less as in staminate flowers; ovary ellipsoid, glabrate, 3-carpellate; ovules numerous, horizontal; styles 3, connate. Fruit green, frequently striped, 2.5–4 cm long, 2.5 cm wide, ovoid to subspherical, glabrous. Seeds ovoid, to 4 mm long, white-sericeous.

Phenology. Flowers in February and October. Fruits in February, May, September, and October.

Distribution. Mexico to Panama. In Costa Rica found from sea level to 1,200 m.

11. GENUS *MOMORDICA*

Our only species of *Momordica* is the introduced *M. charantia*. It is recognized by the following combination of characters. The tendrils are simple, and the eglandular petioles lack tufts of hair at the apex. The leaves are simple, not pinnately lobed, and they lack patelliform glands on leaf undersurfaces near the base. The corolla is yellow with connate petals at least in the lower part. The filaments are distinct, but the anthers are connate. In addition, the pedicels of both staminate and pistillate flowers are generally subtended by a clasping, reniform bract. The fruit is orange, and the seeds are red-arillate.

1. *Momordica charantia* L., Sp. pl. 1009. 1753. Figure 4.8B.

Description. Vine; stems glabrous to tomentose. Leaves simple, to 9 cm long and wide, 5–7-lobate, frequently pubescent below in vein axils; tip acute to acuminate; base cordate. Tendrils simple. Inflorescences solitary, axillary. Staminate flowers subtended by a reniform bract; sepals ovate, to 4 mm long; petals yellow; stamens 3; filaments free; anthers connate. Pistillate flowers with calyx and corolla more or less as in staminate flowers; ovary fusiform, beaked, tuberculate; ovules many, horizontal; stigmas 3, bilobate. Fruit orange, to 6 cm long, ellipsoid or fusiform, tuberculate, beaked, becoming dehiscent. Seeds many, to 1.2 cm long and 0.6 cm wide, red-arillate, oblong, flattened.

Phenology. Flowers from June through January. Fruits in June and from August through February.

Distribution. Probably originally from Africa, but now cultivated throughout Central and South America. In Costa Rica found from sea level to 770 m.

Notes. Well known as *sorosi* in Costa Rica. It reputedly has medicinal value.

12. GENUS *POLYCLATHRA*

In our area, eight genera of the Cucurbitaceae frequently have staminate inflorescences that contain a single flower. *Polyclathra* is distinguished among these by the white corolla, petioles that lack tufts of hair at the apex, and leaves that are neither pinnately lobed nor with patelliform glands below.

1. *Polyclathra cucumerina* Bertol., Novi Comment. Acad. Sci. Inst. Bononiensis 4: 438, t. 36. 1840. Figure 4.8C.

Synonymy. *Pittiera trilobata* Cogn., Bot. Gaz. 20: 289. 1895; *Roseanthus heterophyllus* Brandegee, Univ. Calif. Publ. Bot. 10: 188. 1922.

Description. Vine; stems glabrate to glandular-pubescent. Leaves simple, to 13 cm long and 9 cm wide, 3–5-lobate or angulate, puberulent to glabrate above, pubescent to puberulent below; tip acute to acuminate; base cordate. Tendrils 4-fid, curling only above the bifurcation. Staminate flowers solitary, axillary; sepals linear or linear-subulate, to 1.8 cm long; petals white; stamens 3; filaments free; anthers connate. Pistillate flowers with calyx and corolla more or less as in staminate flowers; ovary ovoid to globose; ovules many, horizontal; style elongated; stigmas 3, bilobate. Fruit ellipsoid to oblong, to 8 cm long, striped, dehiscent. Seeds many, to 6 mm long, ovate, flattened.

Phenology. Flowers from November through March. Fruits from December through March.

Distribution. Mexico to Panama. In Costa Rica found from sea level to 1,000 m.

13. GENUS *PSIGURIA*

Hampshire (1992b) recognized three species of *Psiguria* from Mesoamerica: *P. dunlapii*, *P. triphylla*, and *P. warscewiczii*. The first species is apparently known only from Panama, whereas the latter two have wider distributions including Costa Rica. *Psiguria ovata* was not mentioned by Hampshire (1992b), but was treated as a Costa Rican species by Jeffrey (1978). The group as a whole is in need of thorough revision (Jeffrey 1978). The present treatment tentatively recognizes *P. ovata*, *P. triphylla*, and *P. warscewiczii* from Costa Rica, although only the latter two are found above 700 m.

Identification of our species of *Psiguria* is largely accomplished by leaf thickness and

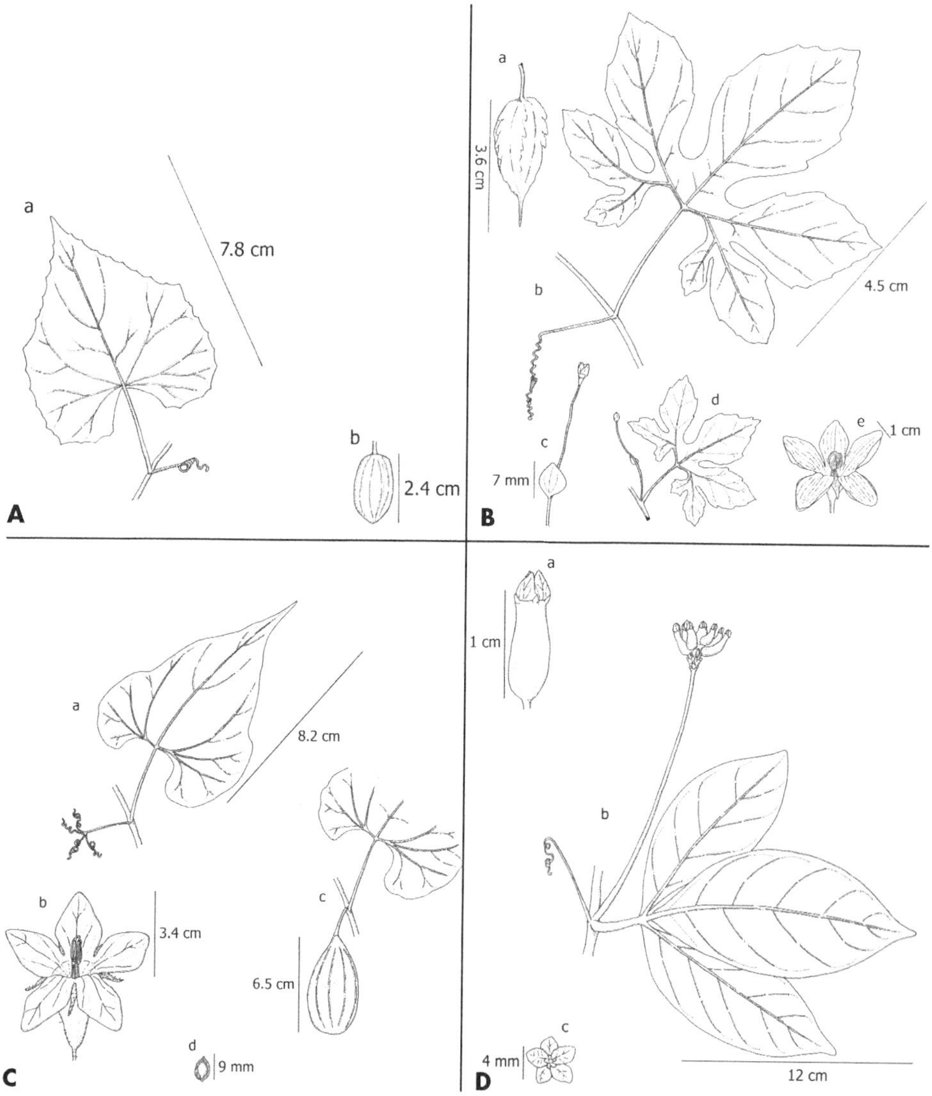

Fig. 4.8. A. *Melothria scabra:* **(a)** leaf and tendril, **(b)** fruit (Based on Taylor 3009, DUKE). **B.** *Momordica charantia:* **(a)** fruit, **(b)** leaf and tendril, **(c)** close-up of inflorescence showing subtending bract, **(d)** leaf and inflorescence, **(e)** staminate flower (Leaves, tendril, and inflorescence based on Dwyer 1006, US; Staminate flower based on Hinton 9800, US; Fruit based on Burch 1514, US). **C.** *Polyclathra cucumerina:* **(a)** leaf and tendril, **(b)** staminate flower, **(c)** leaf and fruit, **(d)** seed (Leaves, tendril, fruit, and seed based on Hammel 7927, DUKE; Staminate flower based on Calderón 2061, US). **D.** *Psiguria triphylla:* **(a)** flower; **(b)** leaf, tendril, and inflorescence; **(c)** corolla (Based on Krings 59, CR).

structure, and color of the calyx tube. Both *P. triphylla* and *P. warscewiczii* usually (but not always) have three foliolate leaves, whereas *P. ovata* generally has simple leaves. *Psiguria triphylla* has thick, fleshy leaves, but both *P. warscewiczii* and *P. ovata* have membranous leaves. Leaf margins of all three species are typically entire, although all three may display lobing to varying degrees. Nonetheless, specimens of *P. triphylla* at US have leaves with less undulate or angulate margins than *P. warscewiczii*. The terminal leaflet in *P. triphylla* is usually wider with a base not so steeply angled as in *P. warscewiczii*. Except for being membranous, leaves of *P. ovata* resemble the simple leaves of *P. triphylla*.

Psiguria triphylla and *P. warscewiczii* have traditionally been distinguished by the color of the calyx tube, dark green-spotted in *P. warscewiczii* and not spotted in *P. triphylla*. But the calyx of the type specimen of *P. ovata* (Tonduz 13006, US) is dark-green spotted as well. However, the leaf bases of both specimens of *P. ovata* seen at US (Tonduz 13006 and Tonduz 11354) are only shallowly cordate, and thus quite different from the deeply cordate bases of some simple-leaved *P. warscewiczii* (e.g., Valerio 26, US). Formerly thought to be dioecious, *Psiguria* is now known to be monoecious.

KEY TO *PSIGURIA*

1. Calyx not green-spotted; leaves generally coriaceous... *P. triphylla*, p. 64.
1. Calyx dark-green spotted; leaves more or less membranous... *P. warscewiczii*, p. 64.

1. *Psiguria triphylla* (Miq.) C. Jeffrey, Kew Bull. 33 (2): 353. 1978. Figure 4.8D.

Synonymy. *Anguria triphylla* Miq., Linnaea 19: 136. 1845; *A. schomburgkiana* Schltdl., Linnaea 24: 765. 1851; *A. pallida* Cogn., Diagn. Cucurb. 1: 22. 1876; *A. diversifolia* Cogn., Bot. Gaz. 16: 10. 1881; *A. elliptica* Britton, Bull. Torrey Bot. Club 50: 54. 1923; *A. vogliana* Suess., Feddes Repert. 30: 278. 1932.

Description. Vine; stems glabrous, sometimes puberulous at nodes. Leaves usually 3-foliolate but sometimes simple, to 20 cm long, somewhat pustulate above, glabrate below; tip acute to acuminate; base cuneate. Tendrils simple. Staminate flowers in racemes, axillary; calyx green, not spotted with dark green, to 1.1 cm long, somewhat tubular, sparsely villous internally, puberulent externally; corolla red to orange, papillose internally and externally; anthers 2, glabrous. Pistillate flowers with calyx and corolla more or less as in staminate flowers; ovary oblong, glabrous. Fruit green, sometimes striped, to 6 cm long, oblong. Seeds white, to 7 mm long and 5 mm wide, ovoid.

Phenology. Flowers from December through July. Fruits in March and June.

Distribution. Mexico to Bolivia. In Costa Rica found from sea level to 1,350 m.

Notes. This species was called *Psiguria bignoniacea* (Poepp. & Endl.) Wunderlin by Wunderlin (1978), as well as in recent lists of Costa Rican species (e.g., Haber 1991; Wilbur 1994; Inbio 1998). In discussing the history of the name, Jeffrey (1978) noted that the type collection of *Anguria bignoniacea* is actually a mixture, consisting of sheets of both a species of *Gurania* and of *Psiguria triphylla*. As the original description apparently applies unequivocally to the former (Jeffrey 1978), Jeffrey applied the basionym of *Anguria bignoniacea* to the species of *Gurania*. The resulting name, *Gurania bignoniacea* (Poepp. & Endl.) C. Jeffrey, belongs to a species found only in South America. Costa Rican material identified as *Psiguria bignoniacea* (Poepp. & Endl.) Wunderlin should be treated as *Psiguria triphylla* (Miq.) C. Jeffrey.

2. *Psiguria warscewiczii* (Hook. f.) Wunderlin, Phytologia 38: 219. 1978. Figure 4.9A.

Synonymy. *Anguria warscewiczii* Hook. f., Bot. Mag. 88, t. 5304. 1862; *A. limonensis* Pittier, Contr. U.S. Natl. Herb. 13: 119, f. 25. 1910; *A. magdalenae* Pittier, Contr. U.S. Natl. Herb. 13: 118, f. 24. 1910; *A. tonduzii* Cogn., in Engl., Pflanzenreich 66 (IV. 275. I.): 191. 1916; *A. aurantiaca* Blake, Contr. U.S. Natl. Herb. 24: 26. 1922.

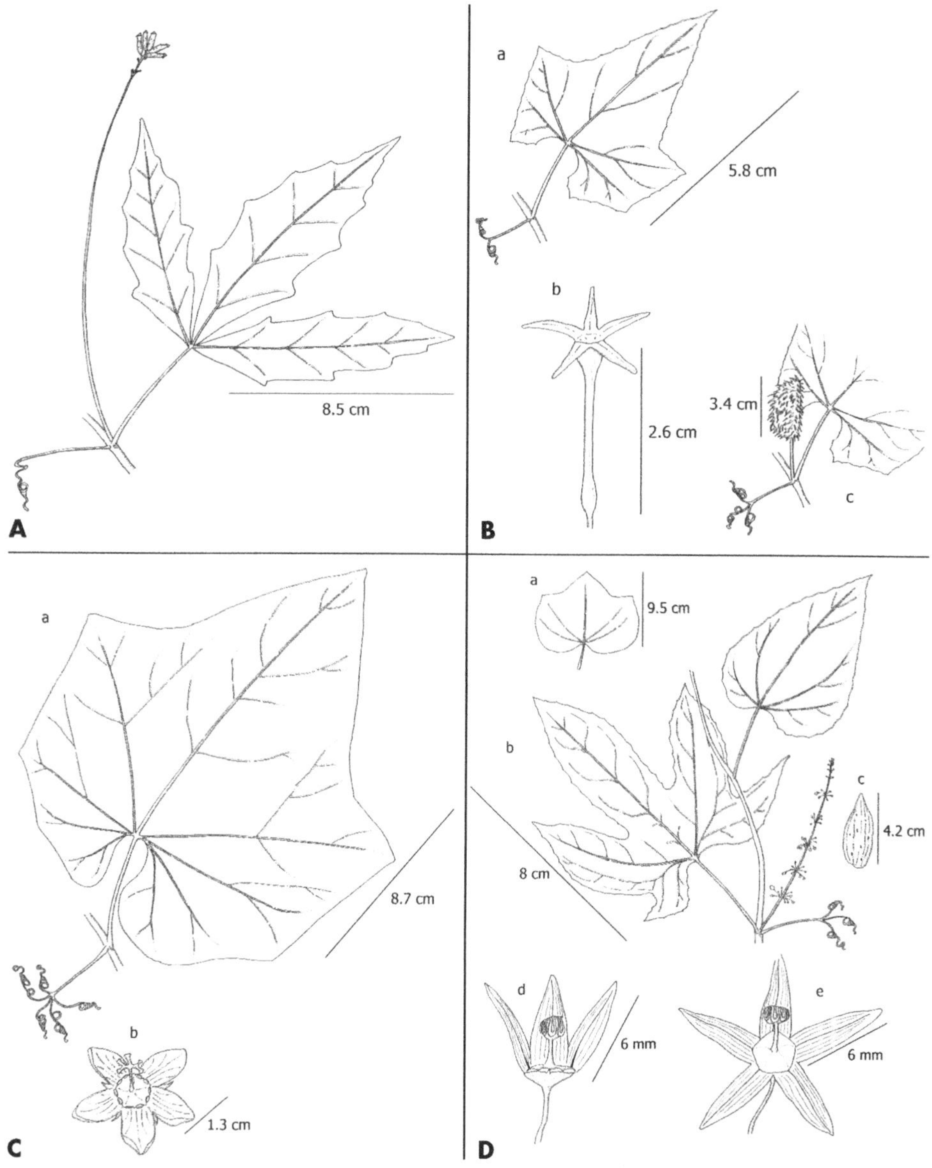

Fig. 4.9. A. *Psiguria warscewiczii:* leaf, tendril, and inflorescence (Based on Tonduz 11536, US). **B.** *Rytidostylis carthaginensis:* **(a)** leaf and tendril; **(b)** staminate flower; **(c)** leaf, tendril, and fruit (Leaves, tendrils, and fruit based on Taylor 338, DUKE; Staminate flower based on Burch 1104, US). **C.** *Sechium edule:* **(a)** leaf and tendril, **(b)** staminate flower (Leaf and tendril based on Gómez 22174, DUKE; Staminate flower based on Krings 278, CR). **D.** *Sechium pittieri:* **(a)** leaf variation; **(b)** leaves, tendril, and inflorescence; **(c)** fruit; **(d)** staminate flower with 2 petals removed; **(e)** staminate flower (Leaves, tendril, inflorescence, and fruit based on Hammel 7833, DUKE; Staminate flowers based on Carlson 3468, F).

Description. Vine; stems glabrate. Leaves 3-foliolate, to 20 cm long, somewhat pustulate above, glabrate or pustulate below; tip acute to acuminate; base cuneate. Tendrils simple. Staminate flowers in racemes, axillary; calyx green, spotted with darker green, to 1.1 cm long, somewhat tubular, glabrate externally, puberulent internally; corolla red to orange, papillose internally and externally; anthers 2, glabrous. Pistillate flowers with calyx and corolla more or less as in staminate flowers; ovary oblong, glabrous. Fruit green, sometimes striped, to 7 cm long, oblong. Seeds white to tannish, to 9 mm long and 6 mm wide, ovoid.

Phenology. Flowers from February through April and from June through October. Fruits in July and October.

Distribution. Belize to Venezuela and Ecuador. In Costa Rica found from sea level to 1,400 m.

14. GENUS *RYTIDOSTYLIS*

Rytidostylis needs additional study and possible revision. In recognizing only one species, *R. carthaginensis*, we followed the conservative treatment of Wunderlin (1978). Future studies may delimit additional species. Our species is distinguished by the rather elongated, narrowly cylindric calyx.

1. ***Rytidostylis carthaginensis*** (Jacq.) Kuntze, Rev. Gen. Pl. 1: 258. 1891. Figure 4.9B.

Synonymy. *Elaterium carthaginense* Jacq., Enum. syst. pl. 31. 1760; *Elaterium heterophyllum* Brandegee, Univ. Calif. Publ. Bot. 4: 387. 1913; *E. ciliatum* Cogn., Mém. Couronnés Acad. Roy. Sci. Belgique 28: 54. 1877; *R. gracilis* Hook. & Arn., Bot. Beechey Voy. 424. 1840–1841.

Description. Vine; stems pubescent to glabrate. Leaves simple, to 11 cm long and 10 cm wide, 3–5-lobate or angled, somewhat scabrous above especially on principal veins, pustulose to puberulent or glabrate above and below; tip acuminate or acute; base cordate. Tendrils 2–3-fid, curling only above the bifurcation. Staminate flowers borne in few to many-flowered umbels or racemes, axillary; calyx tube narrowly elongate-cylindric, to 3 cm long; corolla greenish-white or yellow; corolla lobes papillose internally, glabrous externally; stamens 3; filaments and anthers connate. Pistillate flowers solitary, axillary; calyx and corolla as in staminate flowers; ovary ovoid, echinate; ovules generally ascending; styles and stigmas connate. Fruit ovoid, to 4 cm long and 2 cm wide, echinate, dehiscent. Seeds 12–16, to 1 cm long and 0.8 cm wide, ovate, flattened, winged near base, 3-dentate at apex.

Phenology. Flowers and fruits throughout the year.

Distribution. Mexico to Peru. In Costa Rica found from sea level to 1,750 m.

Notes. This species is a common hedgerow weed in populated areas of Valle Central.

15. GENUS *SECHIUM*

With six species in our area, *Sechium* is recognized by having a single-seeded indehiscent fruit and 10 floral nectaries at the base of the hypanthium. Two sections are recognized based on whether the nectaries are sunken into the hypanthium base (sect. *Sechium*) or formed into a depressed, spherical cushion (sect. *Frantzia*). Our species are evenly divided between the two sections.

Most of our species are found either only in Costa Rica or in both Costa Rica and Panama. Only the widely cultivated *S. edule* is known outside these two countries.

KEY TO *SECHIUM*

1. Floral nectaries sunken into base of hypanthium
   2. Filaments only partially connate; anthers distinct; floral nectaries not protruding below hypanthium or only slightly so... *S. edule*, p. 67.
   2. Filaments completely connate; anthers connate in a head; floral nectaries protruding conspicuously below hypanthium
      3. Floral nectaries glabrous above; leaves glabrous below... *S. tacaco*, p. 68.
      3. Floral nectaries short-tomentose above; leaves pubescent below... *S. talamancensis*, p. 68.
1. Floral nectaries a depressed, more or less spherical cushion on the hypanthium floor
   4. Staminate inflorescence pendulous; staminate flowers subsessile to short pedicellate; pedicels shorter than 3 mm... *S. venosum*, p. 68.
   4. Staminate inflorescence ascending; staminate flowers distinctly pedicellate; pedicels longer than 3 mm
      5. Staminate inflorescence not villous... *S. pittieri*, p. 67.
      5. Staminate inflorescence villous... *S. villosum*, p. 70.

1. ***Sechium edule*** (Jacq.) Sw., Fl. ind. Occid. 1150. 1800. Figure 4.9C.

Synonymy. *Sicyos edulis* Jacq., Enum. Syst. Pl. 32. 1760.

Description. Vine; stems glabrate. Leaves simple, to 18 cm long and wide, 3–5-lobate or angled, somewhat scabrous above, scabrous to glabrate below; tip acute; base cordate. Tendrils 3–5-fid, curling only above the bifurcation. Staminate flowers in racemes; calyx cup-shaped, to 5 mm wide; calyx lobes triangular, to 6 mm long; floral nectaries sunken into base of hypanthium; corolla greenish or yellow; corolla lobes more or less triangular, glabrate to somewhat papillose externally; filaments not connate their entire length, glabrous. Pistillate flowers with calyx and corolla more or less as in staminate flowers, but somewhat smaller; ovary obovoid, glabrous; stigma subglobose. Fruit green to yellow or whitish, to 18 cm long, spiny or smooth. Seed 1.

Phenology. Flowers in August and September. Fruits in August and November.

Distribution. Commonly cultivated in tropical regions of both hemispheres for its edible fruits. In Costa Rica found from sea level to 1,400 m.

Notes. Widely known as *chayote*. Young shoots edible.

2. ***Sechium pittieri*** (Cogn.) C. Jeffrey, Kew Bull. 33 (2): 361. 1978. Figure 4.9D.

Synonymy. *Cyclanthera pittieri* Cogn., Bull. Soc. Bot. Belgique 30: 275. 1891; *Frantzia montana* Pittier, Contr. U.S. Natl. Herb. 13: 128. 1910; *Frantzia pittieri* (Cogn.) Pittier, Contr. U.S. Natl. Herb. 13: 129. 1910.

Description. Vine; stems glabrate. Leaves simple, to 10 cm long and 12 cm wide, 3–5-lobate or angled, glabrous above except for principal veins, pubescent to scabrous below; tip acute; base cordate. Tendrils 3–4-fid, curling only above the bifurcation. Staminate inflorescence racemose-paniculate, not villous, axillary; calyx cup-shaped, to 4 mm wide; calyx lobes subulate, to 1 mm long; floral nectaries formed into depressed, spherical cushion, not sunken into the hypanthium; corolla yellowish, pubescent externally; corolla lobes more or less lanceolate, nearly 4 mm long; anthers in a head, to 2 mm wide. Pistillate flowers with calyx and corolla more or less as in staminate flower. Fruit ovoid, to 4.7 cm long, echinate. Seed 1.

Phenology. Flowers throughout the year. Fruits in January, February, and July.
Distribution. Costa Rica and Panama. In Costa Rica found from sea level to 2,000 m.

3. *Sechium tacaco* (Pittier) C. Jeffrey, Kew Bull. 33: 361. 1978. Figure 4.10A.

Synonymy. *Polakowskia tacaco* Pittier, Contr. U.S. Natl. Herb. 13: 131. 1910; *Frantzia tacaco* (Pittier) Wunderlin, Brittonia 28: 242. 1976.
Description. Vine; stems glabrate. Leaves simple, to 10 cm long and wide, 3–5-lobate or angled, glabrous above and below (sometimes with a few, scattered hairs); tip acute; base cordate. Tendrils 3–5-fid, curling only above the bifurcation. Staminate flowers in racemes, axillary; calyx cup-shaped, to 5 mm wide; calyx lobes subulate, to 1 mm long, glabrous above; floral nectaries sunken into base of hypanthium; corolla greenish or yellow; corolla lobes more or less triangular; filaments connate for their entire length. Pistillate flowers with calyx and corolla more or less as in staminate flowers; ovary somewhat setiferous; stigma capitate. Fruit green, to 6 cm long, somewhat spiny near base. Seed 1.
Phenology. Flowers in February, March, August, and September. Fruits in February, March, August, and September.
Distribution. Endemic to Costa Rica. In Costa Rica found from 1,000 to 1,525 m.
Notes. Fruits known as *tacaco*, and commonly sold in markets. Tender shoots edible.

4. *Sechium talamancensis* (Wunderlin) C. Jeffrey, Kew Bull. 33 (2): 361. 1978. Figure 4.10B.

Synonymy. *Frantzia talamancensis* Wunderlin, Brittonia 28: 242. 1976.
Description. Vine; stems glabrate, sometimes short-tomentose at the nodes. Leaves simple, to 20 cm long and wide, 3–5-lobate or angled, puberulent to pubescent above, pubescent below; tip acuminate; base cordate. Tendrils (3?-)5-fid, curling only above the bifurcation. Staminate flowers in racemes or panicles, axillary; calyx urceolate, to 4 mm wide, glabrous; calyx lobes triangular-lanceolate, to 1 mm long, short-tomentose above; floral nectaries sunken into base of hypanthium; corolla 5-lobed; corolla lobes more or less lanceolate, to 8 mm long; filaments connate their entire length. Pistillate flowers with calyx and corolla more or less as in staminate flowers; ovary fusiform, tomentose to sparsely setiferous; stigma capitate. Fruit green, to 6 cm long, broadly fusiform, somewhat spiny near base. Seed 1.
Phenology. Flowers in July, September, and November. Fruits from August through November.
Distribution. Endemic to Costa Rica. Found from 1,500 to 2,800 m.

5. *Sechium venosum* (L.D. Gómez) Lira & Chiang, Novon 2: 227. 1992. Figure 4.10C.

Synonymy. *Frantzia venosa* L.D. Gómez, Phytologia 53: 447. 1983.
Description. Vine; stems puberulent or glabrous. Leaves simple, to 15 cm long and 12 cm wide, 3–5-lobate or angled, glabrous above and below; tip acute; base cordate. Tendrils 4-fid, curling only above the bifurcation. Staminate flowers in racemes, axillary; calyx shallow, ca. 2.5 mm wide; floral nectaries a depressed, spherical cushion, not sunken into the hypanthium; corolla lobes 5, to 4 mm long, triangular. Pistillate flowers with calyx and corolla more or less as in staminate flower. Fruit oblong to fusiform, echinate. Seed 1.

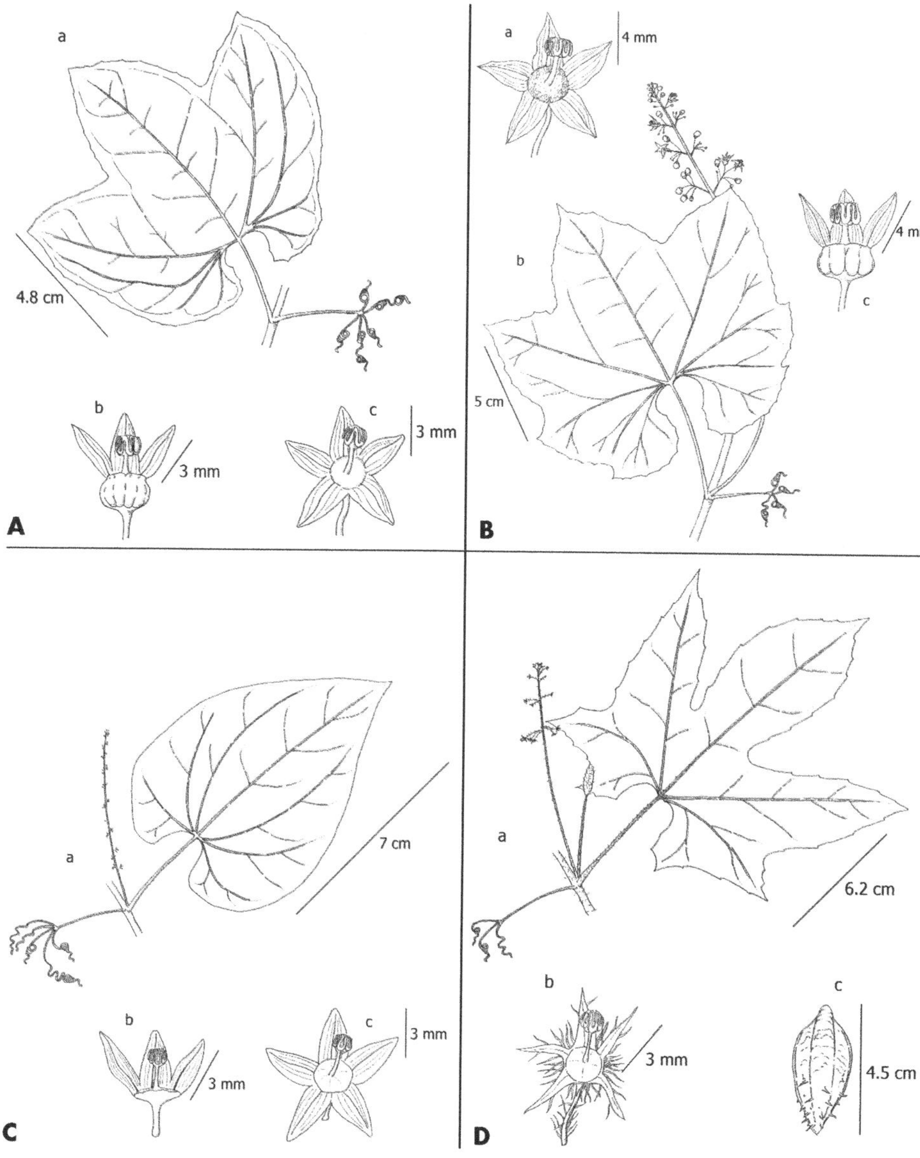

Fig. 4.10. A. *Sechium tacaco:* **(a)** leaf and tendril, **(b)** staminate flower with 2 petals removed, **(c)** staminate flower (Leaf and tendril based on Raven 21752, DUKE; Staminate flowers based on Raven 21859A, F). **B.** *Sechium talamancensis:* **(a)** staminate flower; **(b)** leaf, tendril, and inflorescence; **(c)** staminate flower with 2 petals removed (Leaf, tendril, and inflorescence based on Wilbur 29142, DUKE; Staminate flowers based on Burger and Baker 10052, F). **C.** *Sechium venosum:* **(a)** leaf, tendril, and inflorescence; **(b)** staminate flower with 2 petals removed; **(c)** staminate flower (Leaf, tendril, and inflorescence based on Burger and Matta U. 4317, F; Staminate flowers based on Tonduz 9334, F). **D.** *Sechium villosum:* **(a)** leaf, tendril, inflorescence, and infructescence; **(b)** staminate flower; **(c)** fruit (Leaf, tendril, inflorescence, and infructescence based on Lent 2828, F; Staminate flower and fruit based on Burger and Gentry, Jr. 9227, F).

Phenology. Flowers in February, March, July through September. Fruits in February, March, August, and November.

Distribution. Costa Rica and Panama. In Costa Rica found from sea level to 1,700 m.

6. *Sechium villosum* (Wunderlin) C. Jeffrey, Kew Bull. 33: 361. 1978. Figure 4.10D.

Synonymy. *Frantzia villosa* Wunderlin, Brittonia 28: 240. 1976.

Description. Vine; stems villous, glabrescent in age. Leaves simple, to 25 cm long and wide, 5-lobate or angulate, sparsely villous above and below; tip acuminate; base cordate. Tendrils 5-fid, curling only above the bifurcation. Staminate inflorescence racemose-paniculate, villous, axillary; calyx broadly campanulate, to 7 mm wide; calyx lobes linear-lanceolate, to 8 mm long; floral nectaries a depressed, spherical cushion, not sunken into the hypanthium; corolla lobes 5, to 1 cm long, lanceolate, glabrous internally, villous externally; anthers connate in a head, to 3 mm wide. Pistillate flowers with calyx and corolla more or less as in staminate flower. Fruit ovoid to fusiform, to 4.7 cm long, somewhat spiny at base. Seed 1.

Phenology. Flowers in February, March, and August. Fruits in January.

Distribution. Endemic to Costa Rica. Found from 1,800 to 2,000 m.

16. GENUS *SELYSIA*

Our only species of *Selysia*, *S. prunifera*, is recognized by the large, triangular, arrowhead-shaped seeds. *Selysia* is one of five genera with patelliform glands on the abaxial leaf surface (see also discussion under *Cayaponia*), but it is the only genus having these glands frequently scattered throughout the abaxial leaf surface (not crowded only near the base), and having anthers that are straight or merely slightly curved at the ends.

1. *Selysia prunifera* (Poepp. & Endl.) Cogn. in DC., Monogr. phan. 3: 736. 1881. Figure 4.11A.

Synonymy. *Melothria prunifera* Poepp. & Endl., Nov. Gen. sp. pl. 2: 55. 1838; *Selysia aspera* Killip ex C. Jeffrey, Kew Bull. 25: 235. 1971; *S. perlobata* Killip ex C. Jeffrey, Kew Bull. 25: 235. 1971.

Description. Vine; stems glabrous to pubescent. Leaves simple, 10–21 cm long, 9–17 cm wide, 3-lobate, scattered to densely pubescent above and below on principal veins; veins palmate at base; tip acuminate; base cordate; margins minutely serrulate or denticulate. Tendrils bifid, curling only above the bifurcation. Staminate flowers 1–3, axillary; calyx cup-shaped, 2–4 cm long, 5-lobate, puberulent; corolla white, to 1.5 cm long, campanulate, tomentose internally, puberulent externally; stamens 3, adnate to hypanthium wall but otherwise free, replicate at tip and base. Pistillate flowers solitary, axillary; calyx and corolla as in staminate flowers; ovary subglobose; ovules erect; styles 3, connate; stigmas 3. Fruit red to reddish-orange, to 4 cm wide, subglobose to ellipsoid, indehiscent. Seeds triangular, to 1.5 cm long and 0.8 cm wide, arrowhead-shaped.

Phenology. Flowers in October. Fruits from June through September, in November, and from January through March.

Distribution. Costa Rica to Peru. In Costa Rica found from sea level to 1,100 m.

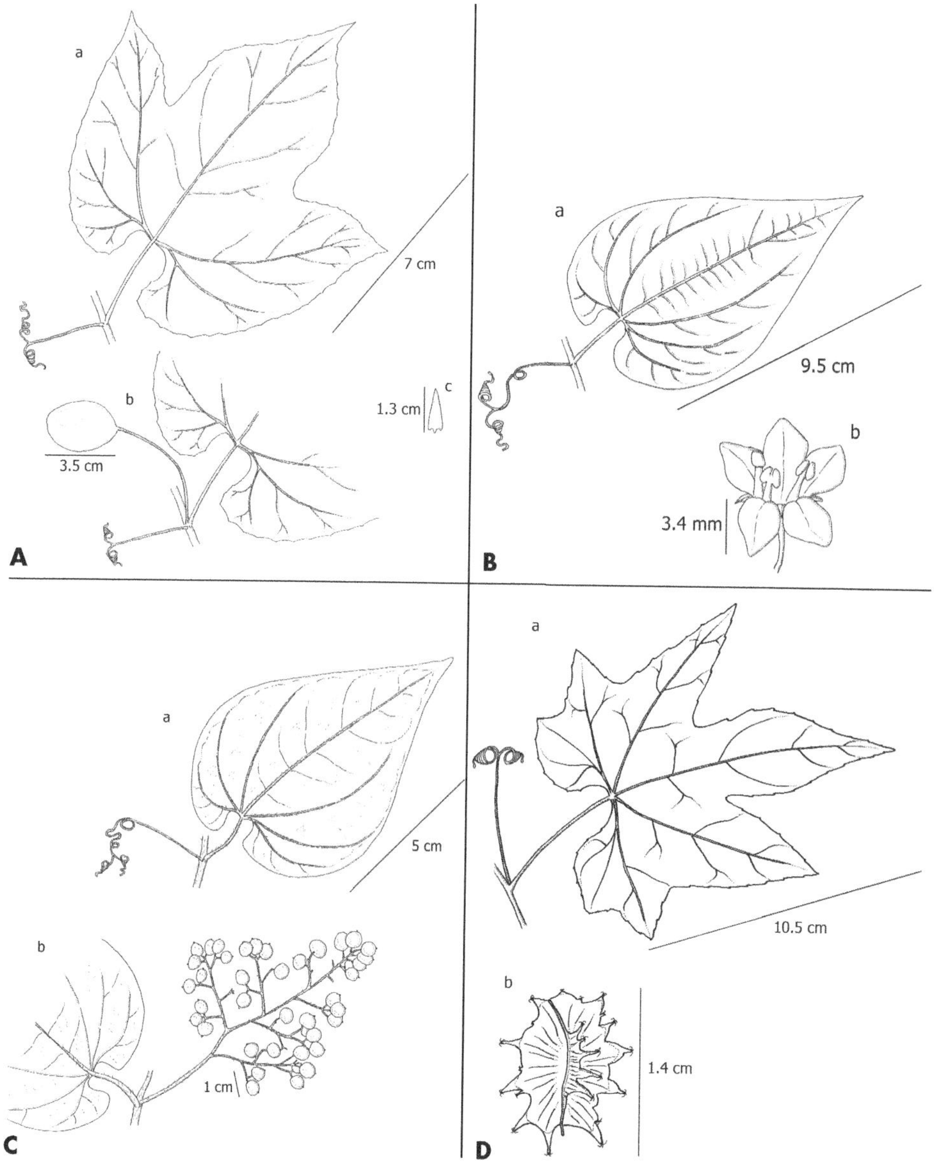

Fig. 4.11. A. *Selysia prunifera:* **(a)** leaf and tendril; **(b)** leaf, tendril, and infructescence; **(c)** seed (Based on Chacón 768, DUKE). **B.** *Sicydium schiedeanum:* **(a)** leaf and tendril, **(b)** staminate flower (Leaf and tendril based on Hansen and Nee 7618, F). **C.** *Sicydium tamnifolium:* **(a)** leaf and tendril, **(b)** leaf and infructescence (Based on Krings 63, CR). **D.** *Sicyos chiriquensis*: **(a)** leaf and tendril, **(b)** fruit (Based on Hammel et al. 6823, DUKE).

17. GENUS *SICYDIUM*

Two species of *Sicydium* occur in our area. The genus can be recognized by the combination of bifid tendrils curling above and below the bifurcation, three stamens, and a unilocular ovary. The only other genus of Cucurbitaceae with branched tendrils that curl on both sides of the bifurcation is *Fevillea* (which has five stamens and a 3-locular ovary). Our treatment of *Sicydium* is tentative. As with other genera in the Cucurbitaceae, additional study is needed. See Lira (1995) for further information.

KEY TO *SICYDIUM*

1. Filaments about as long as the anthers or longer... *S. schiedeanum*, p. 72.
1. Filaments obsolete; anthers sessile or nearly so... *S. tamnifolium*, p. 72.

1. *Sicydium schiedeanum* Schltdl., Linnaea 7: 388. 1832. Figure 4.11B.

Synonymy. *Sicydium glabrum* Standl. & Steyerm., Publ. Field Mus. Nat. Hist., Bot. Ser. 23: 95. 1944.

Description. Vine; stems glabrous to puberulent. Leaves simple, 5–12 cm long, 3–9 cm wide, ovate, puberulent above and below on principal veins; tip acuminate; base cordate; margins entire to somewhat undulate. Tendrils bifid, curling above and below the bifurcation. Inflorescence a many-flowered panicle, axillary. Staminate flowers with sepals lanceolate, less than 1 mm long, villous externally; corolla greenish-white, papillose internally and externally; stamens 3. Pistillate flowers with calyx and corolla as in staminate flowers; ovary narrowly ovoid, to 2 mm long, pubescent; ovule 1, pendulous; styles 3, spreading; stigmas linear. Fruit black, to 1.2 cm wide, globose, indehiscent. Seed 1, 6–8 mm wide, verrucose.

Phenology. Flowers in November. Fruiting times unknown.

Distribution. Mexico to Peru. In Costa Rica found from 1,300 to 1,600 m.

2. *Sicydium tamnifolium* (Kunth) Cogn. in DC., Monogr. phan. 3: 905. 1881. Figure 4.11C.

Synonymy. *Fevillea tamnifolia* Kunth in H.B.K., Nov. Gen. sp. pl. 7: 175, t. 640. 1825; *Bryonia dioica* sensu Sessé & Mocino, Fl. mexic. ed. 2. 1894, non Jacq.

Description. Vine; stems densely pubescent, especially when young. Leaves simple, 6–11 cm long, 4–9 cm wide, ovate, pubescent above and below at least on principal veins; tip acuminate to acute; base cordate; margins entire. Tendrils bifid, curling above and below the bifurcation. Inflorescence a many-flowered panicle, axillary. Staminate flowers with sepals triangular, less than 1 mm long, villous externally; corolla greenish-white, papillose internally and externally; stamens 3. Pistillate flowers with calyx and corolla as in staminate flowers; ovary narrowly ovoid, to 1.5 mm wide, pubescent; ovule 1, pendulous; styles 3; stigmas linear. Fruit black, to 7 mm wide, globose, indehiscent. Seed 1, to 4 mm wide, verrucose.

Phenology. Flowers from June through February. Fruiting times unknown.

Distribution. Mexico to Peru. In Costa Rica found from sea level to 1,700 m.

18. GENUS *SICYOS*

Sicyos is represented by two species in our area. Among the species with white or yellowish flowers in panicles or racemes, and with both connate filaments and anthers, the genus may be distinguished by the salverform corolla, vertically plicate or arched anthers, and lack of nectaries.

KEY TO *SICYOS*

1. Fruits winged, samaralike, bearing recurved barbs at the tips of marginal serrations... *S. chiriquensis*, p. 73.
1. Fruits not winged, but echinate throughout... *S. sertuliferus*, p. 73.

1. *Sicyos chiriquensis* Hammel & D'Arcy, Ann. Missouri Bot. Gard. 68: 215. 1981. Figure 4.11D.

Description. Herbaceous vine; stems glabrous to somewhat pubescent when young. Leaves simple, 5–15 cm long, 5–15 cm wide, 3–7-lobulate, scabrous above, hirsute below; tip acuminate or acute; base cordate. Tendrils bifid, curling only above the bifurcation. Staminate inflorescence a (ca.) 10-flowered raceme, axillary. Staminate flowers 4–6 mm long, 3–4 mm wide, yellow-green, campanulate; calyx 5-lobed; calyx lobes linear, to 1 mm long; stamens 5, united. Pistillate inflorescence capitate. Pistillate flowers smaller than the staminate, ca. 3 mm long, 5-merous, salverform; ovary flat, elliptic, each face longitudinally ridged; ovary ridges toothed with recurved barbs; stigmas 2–3-lobate. Fruits samaroid, ca. 2 cm long, 1.5 cm wide, ridged; fruit ridges longitudinal, serrate with recurved barbs. Seed 1.

Phenology. Flowers and fruits in February.

Distribution. Panama and Costa Rica. In Costa Rica, known from Cartago at 2,060 m (Rodriguez et al. 2006, MO).

2. *Sicyos sertuliferus* Cogn., Bull. Soc. Bot. Belgique 1: 277. 1891. Figure 4.12A.

Description. Vine; stems pubescent. Leaves simple, to 8 cm long and 6 cm wide, 3–5-lobate, pubescent above and below with hairs on the principal veins frequently larger than those between the veins; tip acuminate to acute; base cordate. Tendrils 3–5-fid, curling only above the bifurcation. Inflorescence a many-flowered, long-pedunculate panicle or raceme, axillary. Staminate flowers with calyx more or less campanulate, 5-lobate; corolla yellowish-green (?), salverform; stamens 3; filaments and anthers connate. Pistillate flowers in short-pedunculate clusters; calyx and corolla as in staminate flowers; ovary ovoid, setose to echinate; ovule 1, pendulous; stigmas 2–3-lobate. Fruit trigonous to ovoid, frequently at least 1.2 cm long, dry, echinate. Seed 1.

Phenology. Flowers and fruits in December.

Distribution. El Salvador and Costa Rica. In Costa Rica found from 1,100 to 1,600 m.

19. GENUS *TECUNUMANIA*

Tecunumania quetzalteca is readily identified by the tufts of hair at the apex of the petiole and the (frequently present) sessile, patelliform glands on leaf undersides near the base. The long, linear sepals are also diagnostic.

1. *Tecunumania quetzalteca* Standl. & Steyerm., Publ. Field Mus. Nat. Hist., Bot. Ser. 23: 97. 1944. Figure 4.12B.

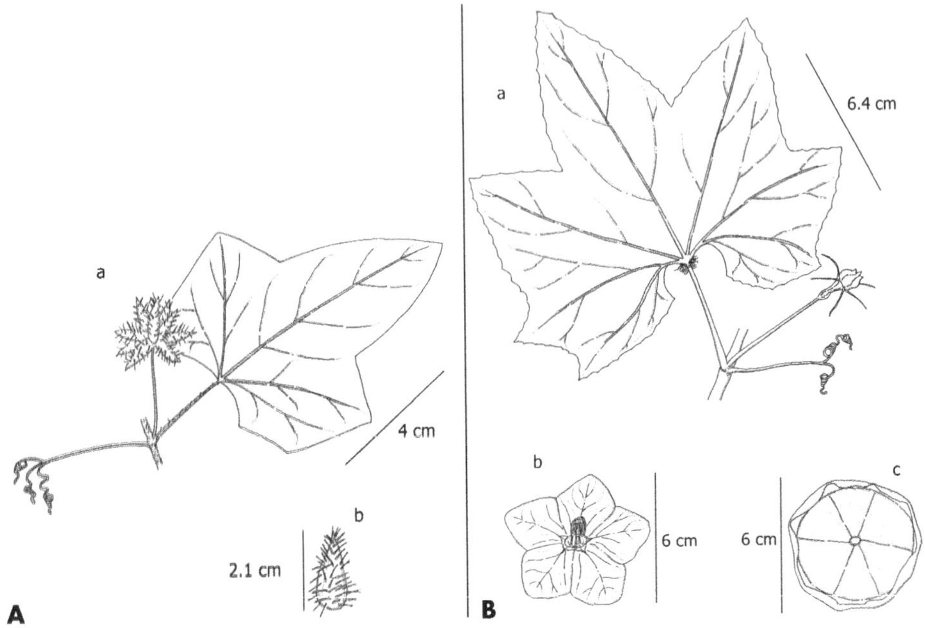

Fig. 4.12. A. *Sicyos sertuliferus:* **(a)** leaf, tendril, and infructescence; **(b)** fruit (Based on Palmer 1176, US). **B.** *Tecunumania quetzalteca:* **(a)** leaf, tendril, and inflorescence; **(b)** staminate flower; **(c)** cross-section of fruit (Based on Primack 413, DUKE).

Description. Vine; stems glabrous, sometimes villous at nodes. Leaves simple, to 17 cm long and 16 cm wide, 3–7-lobate, often with patelliform glands below near leaf base; tip acuminate to acute; base cordate. Tendrils 2–4-fid, curling only above the bifurcation. Inflorescences solitary, axillary. Staminate flowers with sepals linear, to 2 cm long, spreading or reflexed; corolla whitish, deeply 5-lobate; stamens 3, free; anthers united in a head. Pistillate flowers with calyx and corolla as in staminate flowers; stigmas 3, bilobate. Fruit subglobose, to 8 cm long, indehiscent. Seeds whitish, to 8 mm long and 5 mm wide, many, flattened; base rounded.

Phenology. Flowers from October through March. Fruits from January through April.

Distribution. Mexico to Costa Rica. In Costa Rica found from 450 to 1,600 m.

5
FABACEAE

A large family of herbs, shrubs, trees and climbers, the Fabaceae contains about 700 genera and 17,000 species and has a cosmopolitan distribution (Chant 1978). Along with the Bignoniaceae, it is one of the most important liana families of neotropical areas (Gentry 1991). Three subfamilies are generally recognized, Faboideae (or Papilionoideae), Caesalpinioideae, and Mimosoideae. Although several climbing genera occur in Costa Rica (mostly in the Faboideae), only three tendrillate genera are found. Other useful references for this family include Croat (1978), Standley and Steyermark (1946), Woodson and Schery (1950), and Wunderlin (1976).

KEY TO FABACEAE

1. Leaves simple or bifoliolate; tendrils axillary... *Bauhinia*, p. 75.
1. Leaves pinnately or bipinnately compound; tendrils terminating the leaves
   2. Leaves bipinnately compound... *Entada*, p. 76.
   2. Leaves pinnately compound... *Vicia*, p. 77.

1. GENUS *BAUHINIA*

The genus *Bauhinia* comprises trees, shrubs, and both tendrillate and nontendrillate climbers in Costa Rica. The climbing species are typically from lowlands, and only the following species has been collected from our area.

1. *Bauhinia guianensis* Aubl., Hist. pl. Guiane 1: 377. 1775. Figure 5.1A.

Synonymy. *Bauhinia excisa* (Griseb.) Hemsl., Biol. Cent.-Amer., Bot. 1: 337. 1880; *B. manca* Standl., Publ. Field Mus. Nat. Hist., Bot. Ser. 18: 511. 1937.

Description. Liana; stems angled or nearly terete, becoming flattened and sigmoid in age, pubescent or puberulous. Tendrils simple, arising in leaf axils. Leaves frequently bilobed or deeply cleft, to 30 cm long and wide, ovate to ovate-orbicular, sometimes 2-foliolate, glabrate to pubescent above and below; margins entire; petioles to 12 cm long; pulvini somewhat swollen. Inflorescence racemose, many-flowered. Flowers with pedicels to 4 mm long; calyx pubescent, swollen in bud; petals villous externally; stamens 10. Fruit a legume, to 11 cm long and 3.5 cm wide, flattened. Seeds orbicular, to 1.3 cm wide, flattened, notched near hilum scar.

Phenology. Flowers from January through February and from July through October. Fruits in October and November.

Distribution. Mexico to Bolivia, Trinidad. In Costa Rica found from sea level to 800 m.

Notes. Older stems of this species tend to become quite wavy and are unmistakable in the forest. Due to the undulations, the species is commonly called monkey ladder, *bejuco de cadena* (chain vine) or monkey vine, *bejuco de mono*.

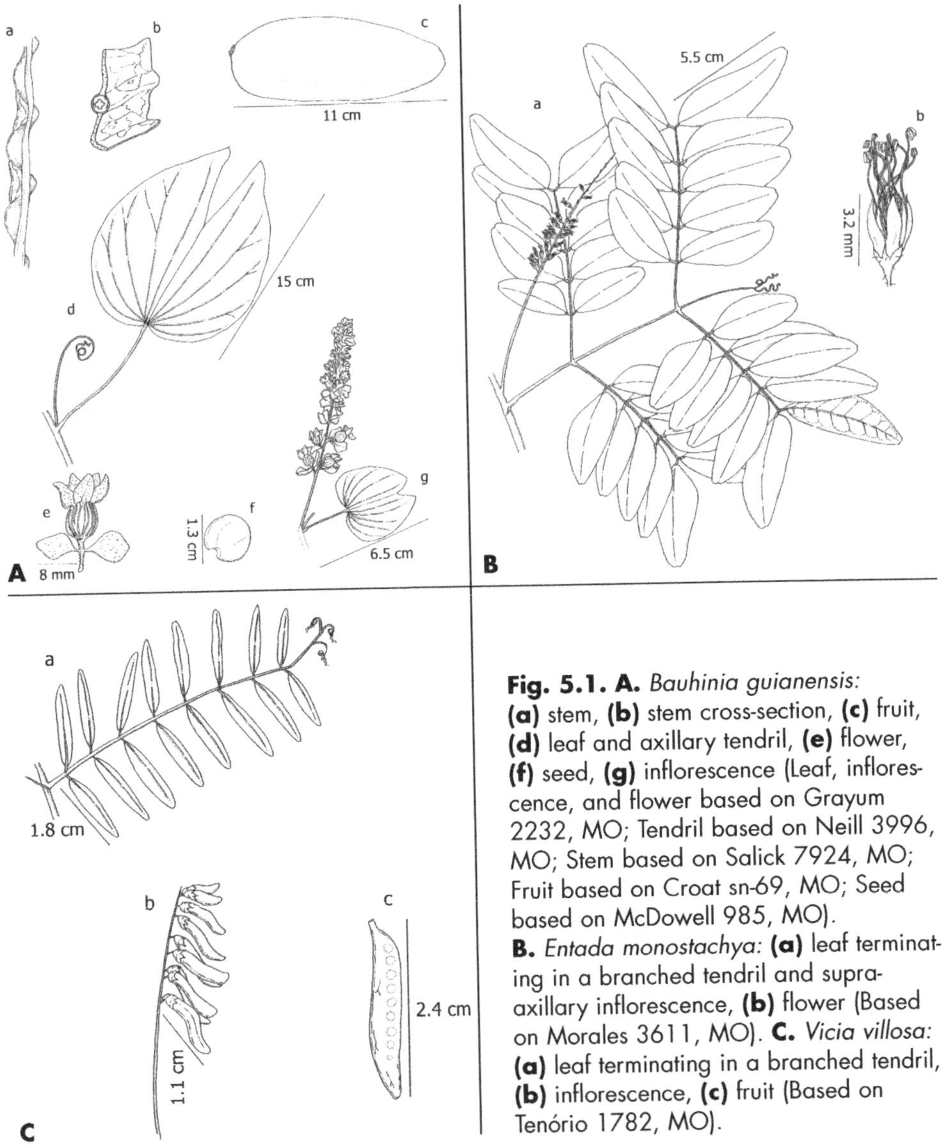

Fig. 5.1. A. *Bauhinia guianensis:* **(a)** stem, **(b)** stem cross-section, **(c)** fruit, **(d)** leaf and axillary tendril, **(e)** flower, **(f)** seed, **(g)** inflorescence (Leaf, inflorescence, and flower based on Grayum 2232, MO; Tendril based on Neill 3996, MO; Stem based on Salick 7924, MO; Fruit based on Croat sn-69, MO; Seed based on McDowell 985, MO). **B.** *Entada monostachya:* **(a)** leaf terminating in a branched tendril and supra-axillary inflorescence, **(b)** flower (Based on Morales 3611, MO). **C.** *Vicia villosa:* **(a)** leaf terminating in a branched tendril, **(b)** inflorescence, **(c)** fruit (Based on Tenório 1782, MO).

2. GENUS *ENTADA*

A genus of trees, shrubs, and climbers, *Entada* is represented by only one tendrillate liana in our area, *E. monostachya*. It is easily identified by bipinnately compound leaves with a rachis that terminates in a branched tendril. No other species in any other family shares these characters. Largely a lowland species, *E. monostachya* is rarely encountered above 700 m. To date, only one collection has been made in our area.

1. *Entada monostachya* DC., Prodr. 2: 425. 1825. Figure 5.1B.

Synonymy. *Entada gigalobium* DC., Prodr. 2: 424. 1825; *E. gigas* (L.) Fawc. & Rendle, Fl. Jamaica 4: 124. 1920.

Description. Liana; stems glabrous. Leaves bipinnately compound, typically 2-jugate, terminating in a branched tendril; leaflets obliquely oblong, to 8 cm long and 3 cm wide, glabrous above and below; tip frequently emarginate; base oblique, obtuse; margins entire; stipules linear, to 4.1 mm long. Inflorescence a spike, supra-axillary, emerging 5–10 mm above the leaf. Flowers creamish, somewhat malodorous; calyx 5-merous, about 1 mm long, glabrous or with scattered hairs; corolla 5-merous; petals elliptic to oblong, to 3 mm long, free or slightly connate basally; stamens 10. Fruit a legume, to 200 cm long and 10 cm wide, oblong, frequently curved, glabrous. Seeds reddish-brown, to 5 cm wide, orbicular, flattened, somewhat glossy.

Phenology. Flowers in April. Fruiting times unknown.

Distribution. Central America and West Indies to northern South America and western Africa. In Costa Rica found from sea level to 900 m.

3. GENUS *VICIA*

The genus *Vicia* is typically recognized by pinnately compound leaves with numerous, small leaflets, and a rachis that terminates in a tendril. Only two other genera also exhibit a rachis terminating in a tendril. Of these two, *Entada* (Fabaceae) is distinguished by its bipinnately compound leaves and *Cobaea* (Polemoniaceae) by its 2–3-jugate leaves (4–8-jugate in *Vicia*).

1. ***Vicia villosa*** Roth, Tent. fl. Germ. 2: 182. 1789. Figure 5.1C.

Description. Vine; stems whitish pilose to villous. Leaves pinnately compound, 4–8-jugate; rachis terminating in a tendril; leaflets linear to oblong, 1–3 cm long, to 8 mm wide, pilose or villous above and below; tip rounded or acute, mucronate; margins entire; stipules linear, to 7 mm long. Inflorescence racemose, densely flowered. Flowers with the calyx somewhat villous; corolla violet to purple. Fruit a legume, to 3 cm long and 1 cm wide, oblong, glabrous. Seeds small, black, globose.

Phenology. Flowers in February and April. Fruits in February and April.

Distribution. Native to Europe, this species is often planted for forage, and has become naturalized in various parts of the world including North and Central America. In Costa Rica found to about 2,300 m.

6
LOGANIACEAE

Worldwide the Loganiaceae contains roughly 30 genera and 600 species of trees, shrubs, and climbers (Jones 1978). In our area the family is represented by seven species of tendrillate lianas, all in the genus *Strychnos*. The genus is easily identified vegetatively by the opposite, simple, 3- or 5-plinerved leaves, and simple, hooklike tendrils. The only other tendrillate family with opposite leaves in our area, the Bignoniaceae, is readily distinguished by compound leaves.

1. GENUS *STRYCHNOS*

Without flowers most of our *Strychnos* cannot be identified to species. *Strychnos toxifera*, perhaps the most readily identified species vegetatively, is characterized by conspicuously hirsute (often rufescent) stems and leaves. *Strychnos* is rich in alkaloids, and at least 19 species contain ingredients used by indigenous Amazonian tribes for curare arrow poisons (Krukoff 1965; Reis and Lipp 1982).

Leeuwenberg (1969) identified three types of tendrils among African *Strychnos*, and suggested their potential taxonomic utility. Although Krukoff and Barneby (1971) concluded that tendril type was of limited taxonomic use in American *Strychnos*, it still provides useful taxonomic information when flowers are lacking. Derived from lateral shoots, tendrils differ mostly in location and number, and whether the associated branchlets contain leaves. Heretofore, only three tendril types have been recognized explicitly (Leeuwenberg 1969; Krukoff and Barneby 1971), but at least four types occur among American *Strychnos* (fig. 6.1

Fig. 6.1. Tendril type among American species of *Strychnos:* **(a)** Type I, **(b)** Type II, **(c)** Type III, **(d)** Type IV.

and table 6.1). Type IV is newly described herein based on a review of specimens at US. Tendrils are often lacking on herbarium specimens, and thus our treatment of tendrils is somewhat tentative. Synonymy follows Krukoff and Monachino (1942). Other important references for this family include Blackwell (1967), Ducke (1955), and Krukoff and Barneby (1969).

Table 6.1. Tendril Types among American Species of *Strychnos*

Tendril	Description
Type I	Solitary in axils of either bracts or leaves.
Type II	Paired in axils of bracts, not leaves; terminal on branchlets forming an anchor-shaped grapple.
Type III	Serial in 2–3 pairs at successive nodes of a leafless branchlet.
Type IV	Paired in opposing leaf axils, not terminal on branchlets.

KEY TO *STRYCHNOS*

1. Inflorescence terminal
    2. Corolla tube not pilose internally, glabrous externally; anthers 2 mm or longer, acuminate at base
        3. Inflorescence sparsely puberulent; style pilose... *S. chlorantha*, p. 80.
        3. Inflorescence glabrous; style glabrous... *S. colombiensis*, p. 82.
    2. Corolla tube short pilose to pilose internally, papillose externally; anthers 1 mm or shorter, rounded at base
        4. Filaments at least twice as long as anthers; stems glabrescent to short pubescent with hairs less than 1 mm... *S. tabascana*, p. 83.
        4. Filaments as long as anthers or nearly so; stems hirsute with hairs to 2 mm long, rufescent... *S. toxifera*, p. 84.
1. Inflorescence axillary
    5. Flowers 5-merous
        6. Inflorescence racemose, 1.5–2 cm long with 5–8 flowers... *S. guianensis*, p. 82.
        6. Inflorescence compound, to 7 cm long with 10–27 flowers... *S. panurensis*, p. 83.
    5. Flowers 4-merous
        7. Inflorescence with 12+ flowers; corolla tube longer than corolla lobes; anthers included; seeds 5 or fewer... *S. darienensis*, p. 82.
        7. Inflorescence with 5–8 flowers; corolla tube as long as corolla lobes or nearly so; anthers distinctly exserted; seeds 1–2... *S. guianensis*, p. 82.

1. ***Strychnos chlorantha*** Progel in Mart., Fl. Bras. 6: 273. 1868. Figure 6.2A.

Description. Liana; stems glabrous. Leaves ovate, ovate-lanceolate, or elliptic, 4–18 cm long, 2–6 cm wide, 3-plinerved, primary veins above sometimes distinctly impressed (at least when dried), glabrous above and below; tip acuminate; base acute to attenuate; margin entire; petioles 6–13 mm long. Tendrils simple, hooklike, axillary, type II. Inflorescence cymose, terminal, sparsely puberulent. Flowers with pedicels to 4 mm long; calyx lobes ovate, to 1 mm long, glabrous but ciliate on margins; corolla glabrous internally and externally, not papillose; corolla tube to 1.5 cm long; corolla lobes to 4 mm long, papillose and glabrous internally, glabrous externally; anthers slightly exserted, to 2.5+ mm long,

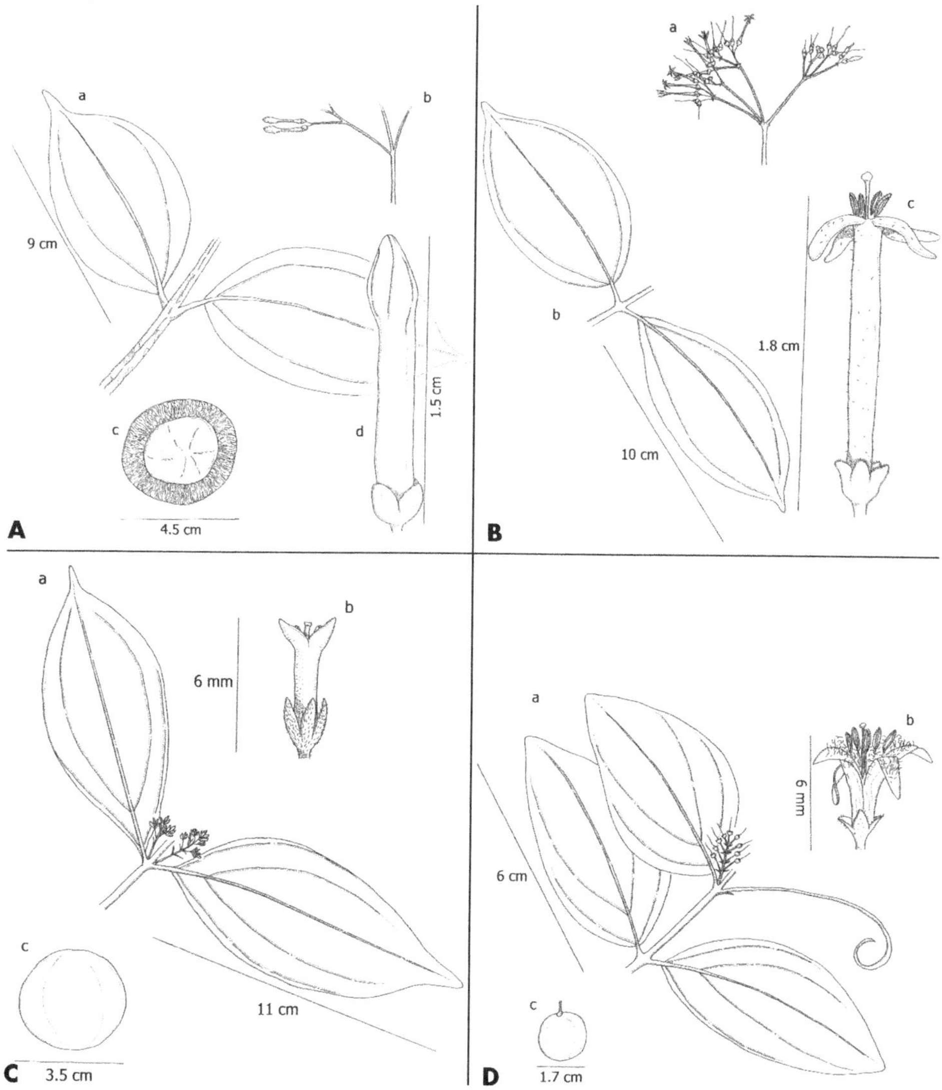

Fig. 6.2. A. *Strychnos chlorantha:* **(a)** leaves, **(b)** inflorescence, **(c)** cross-section of fruit, **(d)** unopened flower (Leaves and fruit based on Barquero M., US; Inflorescence and flower based on Hoffmann 790, US [photo]). **B.** *Strychnos colombiensis:* **(a)** inflorescence, **(b)** leaves, **(c)** flower (Based on Schunke V. 5135, US). **C.** *Strychnos darienensis:* **(a)** leaves and inflorescences, **(b)** flower, **(c)** fruit (Leaves, inflorescences, and flower based on Tonduz 9922, US; Fruit based on Berger s.n., US [photo]). **D.** *Strychnos guianensis:* **(a)** leaves, tendril, and inflorescence; **(b)** flower; **(c)** fruit (Leaves, tendril, and inflorescence based on Cuatrecasas 6921, US; Flower based on Zarucchi 2189, US; Fruit based on Schultes and Cabrera 16473, US).

acuminate and glabrous at base; style pilose. Fruit a capsule, at least 2.5 cm wide, globose, green; shell about 7 mm thick. Seeds about 12.

Phenology. Flowers in February, March, June, October, and November. Fruits in February and March.

Distribution. Guatemala and Costa Rica, but according to Krukoff (1965) also in El Salvador, Honduras, and Nicaragua. In Costa Rica found from about 200 to 1,100 m.

2. *Strychnos colombiensis* Krukoff & Barneby, Mem. New York Bot. Gard. 12: 21. 1965. Figure 6.2B.

Description. Liana; stems cinerous. Leaves ovate to elliptic or elliptic-obovate, 6–10 cm long, 3–6.5 cm wide, 3-plinerved, glabrous above and below; tip short acuminate; base subacute to obtuse; margin entire; petiole to 10 mm long. Tendrils simple, hooklike, axillary. Inflorescence cymose, terminal. Calyx lobes obovate, to 1 mm long, glabrous but ciliate on margins; corolla glabrous internally and externally; corolla tube to 1.8 cm long; corolla lobes to 5.5 mm long, papillose and glabrous internally (except for scattered hairs near sinuses), glabrous externally; stamen 1–3, somewhat excluded; staminodia 2–3, sometimes 1+ intermediate in form; style glabrous. Fruit a capsule, to 6 cm wide, globose, nearly black; shell about 7 mm thick. Seeds discoid, 5 or more.
Phenology. Flowers in February and June. Fruits in April and July.
Distribution. Costa Rica to Peru. In Costa Rica found from 100 to 1,500 m.

3. *Strychnos darienensis* Seem., Bot. voy. Herald 166, 1854. Figure 6.2C.

Synonymy. *Strychnos manaoensis* Barb. Rodr., Vellosia, ed. 2, 1: 36, pl. 5. 1891 (?); *S. ulei* Gilg ex Ule, Bot. Jahrb. Syst. 40: 136. 1907, nomen nudum; *S. macroura* Gilg ex J.F. Macbr., Candollea 5: 400. 1934, nomen nudum (?).
Description. Liana; stems minutely puberulent to glabrous. Leaves ovate, ovate-lanceolate or elliptic, 4.5–18 cm long, 2–8 cm wide, puberulent to glabrous above and below, 3(5)-plinerved, frequently rufescent barbate below in axils of principal veins; primary veins distinctly impressed above (at least when dried); tip acuminate; base acute, subcordate, or rounded; margins entire; petioles to 7 mm long. Tendrils simple, hooklike, axillary, type I. Inflorescence short-racemose or thyrsoid, axillary. Flowers with pedicels to 2 mm long; calyx lobes narrowly lanceolate, to 1.5 mm long, glabrous to hirsutulous externally, ciliate on margins; corolla yellow or white (?), minutely papillose externally; corolla lobes mostly glabrous but bearded internally at base; anthers subsessile; style glabrous. Fruit a capsule, to 4 cm wide, oblong to globose, yellow; shell about 0.7 mm thick. Seeds more or less discoid, fewer than 5.
Phenology. Flowers in March. Fruits in February, June, and September.
Distribution. Nicaragua to Brazil and Bolivia. In Costa Rica found from sea level to 800 m.
Notes. Known as *madronito* in Nicaragua.

4. *Strychnos guianensis* (Aubl.) Mart., Syst. mat. med. bras. 121. 1843. Figure 6.2D.

Synonymy. *Rouhamon guianensis* Aubl., Pl. Guian. 93. 1775; *Toxicaria americana* Schreb., Naturforscher 19: 146. 1783; *Lasiostoma rouhamon* J.F. Gmel., Syst. nat. 2: 248. 1791; *L. cirrhosa* Willd., Sp. pl. 1: 624. 1798; *L. divaricata* G. Mey., Prim. fl. esseq. 83. 1818; *L. curare* Kunth in H.B.K., Nov. Gen. sp. pl. 7: 210. 1825 (?); *Rouhamon divaricatum* DC., Prodr. 9: 17. 1845; *R. guianensis* var. *ecirrhosum* DC., Prodr. 9: 17. 1845; *R. curare* DC., Prodr. 9: 17. 1845 (?); *Strychnos rouhamon* Benth., J. Linn. Soc., Bot. 1: 106. 1856; *S. lanceolata* Spruce ex Benth., J. Linn. Soc., Bot. 1: 105. 1856; *S. curare* Kunth ex Benth., J. Linn. Soc., Bot. 1: 108. 1856 (?); *S. glabra* Sagot ex Progel in Mart., Fl. Bras. 6: 275. 1868 (?); *Rouhamon glabrum* Sagot ex Progel in Mart., Fl. Bras. 6: 275, pro syn. 1868; *Strychnos crevauxiana* Baill., Adansonia 12: 377. 1879; *S. geoffraeana* Baill., Adansonia 12: 378. 1879; *S. crevauxii* G. Planch., Compt. Rend. Acad. Sci. Paris 88: 134. 1880; *S. rivularia* Barb. Rodr., Vellosia, ed. 2, 1: 35. pl. 2, f. B; pl. 4, f. C. 1891 (?); *S. papilosa*

Barb. Rodr., Vellosia, ed. 2, 1: 36. pl. 4, f. B. 1891 (?); *S. oblonga* Gilg, Bot. Jahrb. Syst. 25 Beibl. 60: 37. 1898; *S. calantha* Gilg ex Ule, Bot. Jahrb. Syst. 40: 138, 1907, nomen nudum; *S. guianensis* var. *trichina* J.F. Macbr., Candollea 5: 401. 1934.

Description. Liana; stems minutely but densely pubescent. Leaves ovate, obovate, ovate-lanceolate, or obovate-lanceolate, 3–10 cm long, 1.5–5 cm wide, minutely puberulent above and below; 3(5)-plinerved; tip rounded to acuminate; base cuneate to subcordate; margins entire; petioles to 0.6 cm long. Tendrils simple, hooklike, axillary. Inflorescence in short-racemose cymes, axillary. Flowers with pedicels 4 mm long; calyx lobes ovate to narrowly lanceolate, variously ciliate; calyx lobes to 1.5 mm long; corolla tube to 2.5 cm long, papillose internally and becoming glabrous near upper half (some scattered hairs may remain), woolly internally on lower half, minutely papillose externally; anthers exserted, 0.7–1 mm long; style glabrous. Fruit a capsule, 1.5–2 cm wide, ovate to oblong, orange-yellow; shell to 0.5 mm thick. Seeds more or less discoid, 1–2.

Phenology. Flowering times unknown. Fruits in November.

Distribution. Costa Rica to Bolivia. In Costa Rica found from sea level to 700 m.

5. *Strychnos panurensis* Sprague & Sandwith, Kew Bull. 1927: 132. 1927. Figure 6.3A.

Synonymy. *Strychnos placida* J.F. Macbr., Candollea 5: 400. 1934.

Description. Liana; stems glabrous to distally hirsutulous. Leaves ovate, ovate-lanceolate, or elliptic, 4–20 cm long, 2–8 cm wide, minutely puberulent to glabrous above and below, 3(5)-plinerved; primary veins sometimes distinctly impressed above (at least when dried); tip acuminate; base acute, cuneate, or rounded; margins entire; petioles 3–14 mm long. Tendrils simple, hooklike, axillary, type I. Inflorescence elongate thyrsoid, to 6.5+ cm long, axillary; main axis with several pairs of opposite flowering minor axes. Flowers with pedicels 0.9–4.5 mm long; calyx lobes deltoid, rhombic, or ovate, to 1.5 mm long, glabrous, papillose, or hirsutulous externally, ciliate on margins; corolla yellow or white, salverform, pilose internally, mostly glabrous externally but sometimes minutely papillose; corolla tube 1.5–4 mm long; corolla lobes 2–5.5 mm long; anthers exserted, 1.5–2 mm long, glabrous, rounded at base; style glabrous. Fruit a capsule, to 2.5 cm wide, ovoid to subglobose, yellow or yellowish-green; shell to 0.5 mm thick. Seeds more or less discoid, 1–2.

Phenology. Flowers in February and November in Panama, but flowering times in Costa Rica unknown. Fruiting times unknown.

Distribution. Belize to Peru. In Costa Rica found from sea level to 1,100 m.

6. *Strychnos tabascana* Sprague & Sandwith, Kew Bull. 1927: 128. 1927. Figure 6.3B.

Synonymy. *Strychnos panamensis* var. *hirtiflora* Standl., Publ. Field Mus. Nat. Hist., Bot. Ser. 11: 138. 1932; *S. hirtiflora* Lundell, Bull. Torrey Bot. Club 64: 556. 1937.

Description. Liana or scandent shrub; stems sparsely to scattered pilose; trichomes less than 1 mm long. Leaves ovate, ovate-lanceolate, or elliptic, 4–12 cm long, 1.4–5 cm wide, 3-veined, glabrous or sparsely to densely pilose along principal veins; tip acute to acuminate; base acute to cordate; margins entire; petioles to 4 mm long. Tendrils simple, hooklike, axillary, types I and IV. Inflorescence cymose, terminal. Flowers with pedicels 3.5 mm long; calyx lobes narrowly lanceolate, to 4 mm long; corolla salverform, greenish-yellow or white; corolla tube to 2.5 cm long, dense or scattered, spreading hairs of various

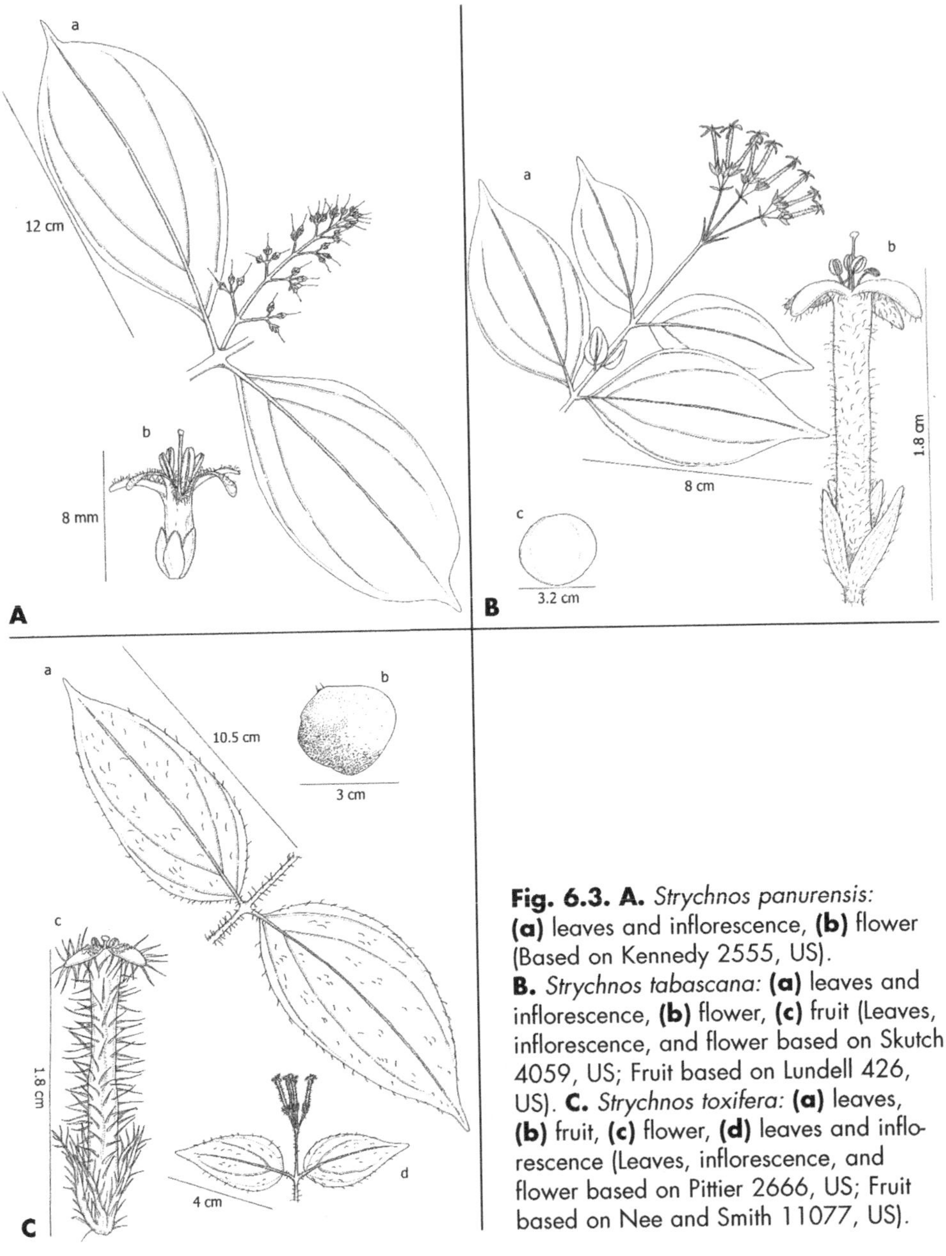

Fig. 6.3. A. *Strychnos panurensis:* **(a)** leaves and inflorescence, **(b)** flower (Based on Kennedy 2555, US). **B.** *Strychnos tabascana:* **(a)** leaves and inflorescence, **(b)** flower, **(c)** fruit (Leaves, inflorescence, and flower based on Skutch 4059, US; Fruit based on Lundell 426, US). **C.** *Strychnos toxifera:* **(a)** leaves, **(b)** fruit, **(c)** flower, **(d)** leaves and inflorescence (Leaves, inflorescence, and flower based on Pittier 2666, US; Fruit based on Nee and Smith 11077, US).

lengths externally; anthers exserted, 0.7–1 mm long. Fruit a capsule, to 8 cm wide. Seeds more than 5.

Phenology. Flowers from December to March, and in July. Fruiting times unknown.

Distribution. Mexico to Costa Rica. In Costa Rica found from 200 to 1,800 m.

7. *Strychnos toxifera* Schomb. ex Benth., J. Bot. (Hooker) 3: 240. 1841. Figure 6.3C.

Synonymy. *Strychnos toxifera* var. *acuminata* Klotzsch in Mart., Fl. Bras. 6: 271. 1868; *S. toxifera* var. *obliqua* Klotzsch in Mart., Fl. Bras. 6: 271. 1868; *S. toxifera* var. *latifolia*

Klotzsch in Mart., Fl. Bras. 6: 271. 1868; *S. syntoxica* Sprague and Sandwith, Kew. Bull. 1927: 129. 1927.

Description. Liana or shrub; stems rufescent-hirsute, trichomes to 2 mm long. Leaves ovate, ovate-lanceolate, or elliptical, 3.5–18 cm long, 2–8.5 cm wide, sparsely to distinctly pubescent above and below, (3-)5-plinerved; tip acuminate; base acute, subcordate, or rounded; margins entire; petioles to 5 mm long. Tendrils simple, hooklike, axillary, often densely rufescent hirsute, type I. Inflorescence cymose, terminal. Flowers with hirsute pedicels to 6 mm long; calyx lobes narrowly lanceolate, 1–7 mm long; corolla salverform, yellow or white; corolla tube to 1.7 cm long, hirsute externally, bearded at base; anthers inserted (tips may protrude), 0.7–1 mm long. Fruit a capsule, to 7 cm wide, globose, green to bluish-green; shell 2.5 mm thick. Seeds more or less discoid, 10–15.

Phenology. Flowers in June. Fruits in February.

Distribution. Costa Rica to Ecuador. In Costa Rica found from 150 to 1,500 m.

7
PASSIFLORACEAE

A mostly tropical and subtropical family of 18 genera and 630 species (Holm-Nielsen et al. 1988), the Passifloraceae is represented in our area only by the genus *Passiflora.* Characterized by an androgynophore and corona, *Passiflora* is unmistakable in flower (fig. 7.1). Vegetatively, the genus is easily identified in our area by alternate leaves and simple, axillary tendrils. The only species lacking tendrils, *P. tica*, is a shrub or small tree.

Bauhinia (Fabaceae) and *Gouania* (Rhamnaceae) are most similar vegetatively to *Passiflora. Bauhinia* is easily distinguished by having arcuate venation, swollen pulvini, and bifoliolate or cleft leaves. *Gouania* is distinguished by having short branches with tendrils subtended by leaves. Various gourds (Cucurbitaceae) may have simple tendrils as well, but these are generally borne at right angles to the petiole. The only other family with alternate leaves and axillary tendrils is the Sapindaceae, easily distinguished by bifid tendrils, compound leaves, and usually white latex.

Although not found in all *Passiflora*, the combination of simple, axillary tendrils and petiole glands always identifies *Passiflora* in our area. The presence of glandular-ocellae on abaxial leaf surfaces is also unique to *Passiflora* in our area. Only seven of our species exhibit neither glandular-ocellae nor glandular petioles.

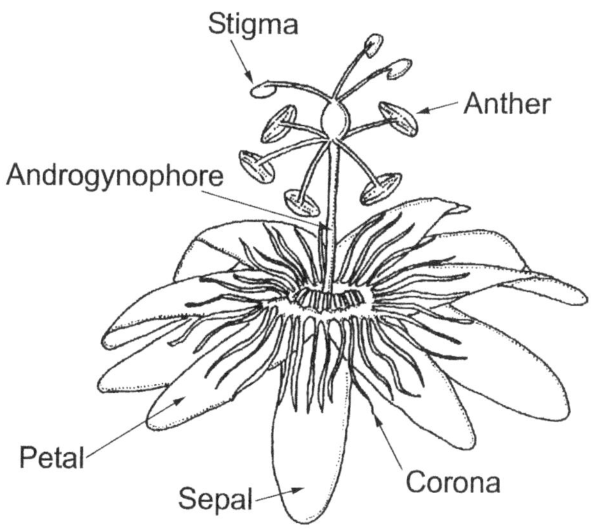

Fig. 7.1. Flower of *Passiflora.*

1. GENUS *PASSIFLORA*

The genus *Passiflora* contains 38 species in our area. Four species are endemic to Costa Rica: *P. boenderi*, *P. dioscoreifolia*, *P. gilbertiana*, and *P. nubicola*. A few species, such as *P. quadrangularis*, are suspected to be native to South America, but are seen mostly in cultivation in our area. Many species have edible fruits, and *P. edulis*, *P. ligularis*, and *P. quadrangularis* are cultivated for this purpose.

Most *Passiflora* are rather easily identified vegetatively, although flowers are needed for some species. Important characters include petiolar glands (presence, absence, location, and number), glandular-ocellae (presence, absence, and location on leaf surface below), stipules and bracts (presence, absence, shape, and size), and floral characters.

Owing to the large number of species, the key to *Passiflora* has been subdivided into four parts (subkeys). For expediency, petiolar glands are considered to occur opposite each other in pairs, although they frequently may actually be subopposite or alternate. Leaf length is measured from the base of the blade to the apex of the midrib. Leaf width is measured across the blade at the widest part. Synonymy follows Killip (1938).

The list and table 7.1 below may further assist with identification. The following species have corniculate sepals: *P. adenopoda*, *P. ambigua*, *P. coccinea*, *P. menispermifolia*, *P. mollissima*, *P. nelsonii*, *P. oerstedii*, *P. quadrangularis*, *P. serratifolia*, and *P. vitifolia*. Table 7.1 lists our species by shape of the stipules.

Table 7.1. *Passiflora* Stipule Shapes

Shape	Species
Narrowly oblanceolate, linear, setaceous, filiform, or subulate	*P. ambigua*
	P. apetala
	P. bicornis
	P. biflora
	P. boenderi
	P. brevifila
	P. capsularis
	P. coccinea
	P. coriacea
	P. costaricensis
	P. edulis
	P. eueidipabulum
	P. gilbertiana
	P. gracilis
	P. helleri
	P. holosericea
	P. lancearia
	P. nelsonii
	P. nubicola
	P. obovata
	P. ornithoura
	P. pusilla
	P. serratifolia
	P. sexflora
	P. standleyi
	P. suberosa
	P. talamancensis
	P. vitifolia
Ovate, lanceolate, reniform, or suborbicular	*P. adenopoda*
	P. dioscoreifolia
	P. guatemalensis
	P. hahnii
	P. ligularis
	P. lobata
	P. membranacea
	P. menispermifolia
	P. mollissima
	P. oerstedii
	P. quadrangularis

KEY TO *PASSIFLORA*

1. Plants lacking both leaf and petiolar glands... Subkey A.
1. Plants with leaf or petiolar glands or both
    2. Plants with petiolar glands only... Subkey B.
    2. Plants with glands only on leaf surfaces below, or on both petioles and leaf surfaces below
        3. Plants with glands only on leaf surfaces below... Subkey C.
        3. Plants with glands on both petioles and leaf surfaces below... Subkey D.

Subkey A: Plants lacking both leaf and petiolar glands

1. Leaves peltate
    2. Stipules setiferously dentate; leaves ovate-orbicular... *P. guatemalensis*, p. 102.
    2. Stipules minutely crenate; leaves suborbicular
        3. Longer peduncles 2–7.5 cm long; some leaves with glands on basal margin... *P. hahnii*, p. 102.
        3. Longer peduncles 6–14 cm long; leaves lacking glands on basal margin... *P. membranacea*, p. 107.
1. Leaves not peltate
    4. Leaves 1.5 times longer than wide
        5. Lateral leaf lobes nearly half or more as long as the midrib... *P. capsularis*, p. 97.
        5. Lateral leaf lobes one-third as long as the midrib... *P. costaricensis*, p. 98.
    4. Leaves about as long as wide, or wider than long
        6. Floral bracts lacking; ovary puberulent (sometimes minutely so), hexagonal or six-ridged in cross-section; fruit fusiform, hexagonal, at least 3 cm long; plant to 0.5 m tall or long... *P. pusilla*, p. 112.
        6. Floral bracts present; ovary densely tomentose to villous, not hexagonal or six-ridged; fruit globose, 1 cm long; plant at least 2 m tall or long... *P. sexflora*, p. 113.

Subkey B: Plants with petiolar glands only

1. Leaves deeply 3- or 5-lobate; apical lobe as large or larger than lateral lobes
    2. Stipules linear, setaceous, or subulate
        3. Glands in lower half of petiole... *P. vitifolia*, p. 116.
        3. Glands in upper half of petiole
            4. Leaf margins serrate... *P. edulis*, p. 99.
            4. Leaf margins entire... *P. suberosa*, p. 115.
    2. Stipules ovate, reniform, or semiorbicular
        5. Petioles with 1 pair of glands (near petiolar apex)... *P. adenopoda*, p. 92.
        5. Petioles with 2 or more pairs of glands
            7. Leaves distinctly serrate... *P. mollissima*, p. 108.
            7. Leaves entire or inconspicuously serrulate
                8. Plant pubescent to puberulent; hairs hooked, ca. 0.5 mm long; stipules 0.5–1.5 cm long... *P. lobata*, p. 107.
                8. Plant densely white-villous; hairs straight, ca. 1.2 mm long; stipules 1.5–3.5 cm long... *P. menispermifolia*, p. 107.
1. Leaves elobate, deeply 2-lobate, or shallowly 3-lobate
    9. Leaves peltate... *P. eueidipabulum*, p. 100.
    9. Leaves not peltate

10. Leaf base cordate
    11. Petioles with 1 pair of glands... *P. dioscoreifolia*, p. 98.
    11. Petioles with 2 or more pairs of glands
        12. Stems sharply tetragonal; angles frequently winged... *P. quadrangularis*, p. 112.
        12. Stems terete to subangulate or angulate, but neither sharply tetragonal nor winged
            13. Petiole with 2 gland pairs; glands subglobose, 1.7 mm long, 3 mm wide... *P. nelsonii*, p. 108.
            13. Petiole with 3 or more gland pairs (rarely 2); glands linear to filiform, 1–10 mm long, 1 mm or less wide
                14. Stipular venation mostly parallel; petiolar glands 3–10 mm long... *P. ligularis*, p. 105.
                14. Stipular venation reticulate, or at least not parallel; petiolar glands 1–2 mm long... *P. oerstedii*, p. 110.
10. Leaf base not cordate
    15. Stems sharply tetragonal; angles frequently winged... *P. quadrangularis*, p. 112.
    15. Stems terete to subangulate or angulate; angles not winged
        16. Petioles with 1 pair of glands
            17. Floral bracts greater than 3 cm long... *P. ambigua*, p. 93.
            17. Floral bracts less than 4 mm long... *P. obovata*, p. 110.
        16. Petioles with 2 or more pairs of glands
            18. Petiolar glands to 1.7 mm long and 3 mm wide... *P. nelsonii*, p. 108.
            18. Petiolar glands to 2 mm long and less than 1 mm wide
                19. Stipules ovate to reniform... *P. oerstedii*, p. 110.
                19. Stipules linear to narrowly, elongate triangular
                    20. Leaves glabrous below; margins entire... *P. brevifila*, p. 96.
                    20. Leaves densely hirtellous below; margins... *P. serratifolia*, p. 113.

Subkey C: Plants with glands only on leaf surfaces below

1. Leaves unlobed to shallowly 3-lobed, clearly longer than wide
    2. Leaves at most 1.5 times longer than wide
        3. Staminal filaments ca. 1 cm or longer; ovary patchily pilose or short-tomentose (surface of ovary easily visible in patches)... *P. helleri*, p. 103.
        3. Staminal filaments to 7 mm or less; ovary densely woolly-pilose or short-tomentose (surface of ovary not visible for the hairs)... *P. talamancensis*, p. 115.
    2. Leaves twice longer than wide
        4. Stipules 1–3 mm long; leaf lobes rounded... *P. lancearia*, p. 105.
        4. Stipules 3.5–6.5 mm long; leaf lobes acute... *P. talamancensis*, p. 115.
1. Leaves strongly 2-lobed or if shallowly 3-lobed, then leaves as wide as long or wider than long
    5. Leaves 1.5 times longer than wide
        6. Leaves not variegated... *P. talamancensis*, p. 115.
        6. Leaves variegated
            7. Lateral leaf lobes more than half the length of the midrib; flowers not malodorous... *P. boenderi*, p. 96.

7. Lateral leaf lobes less than a quarter of the length of the midrib; flowers malodorous... *P. gilbertiana*, p. 100.

5. Leaves about as long as wide, or wider than long

8. Lateral leaf lobes only slightly, if at all, longer than apical lobe... *P. nubicola*, p. 108.

8. Lateral leaf lobes clearly longer than apical lobe

9. Floral bracts 1–1.5 cm long, lanceolate to orbicular-cordate, involucrate... *P. bicornis*, p. 94.

9. Floral bracts absent, or if present shorter than 6 mm, linear, subulate, or setaceous, not involucrate

10. Petals lacking or vestigial... *P. apetala*, p. 94.

10. Petals present, well developed

11. Peduncle 30–40 mm long... *P. standleyi*, p. 115.

11. Peduncle 7–20 mm long

12. Leaves coriaceous, not variegated... *P. biflora*, p. 94.

12. Leaves membranous, variegated... *P. ornithoura*, p. 112.

Subkey D: Plants with glands on both petioles and leaf surfaces below

1. Leaves peltate... *P. coriacea*, p. 97.

1. Leaves not peltate

2. Leaves as long as wide, or wider than long, all three lobes more or less equal in size... *P. gracilis*, p. 102.

2. Leaves longer than wide; apical lobe conspicuously larger than the lateral lobes... *P. holosericea*, p. 103.

1. ***Passiflora adenopoda*** DC., Prodr. 3: 330. 1828. Figure 7.2A.

Synonymy. *Passiflora acerifolia* Schltdl. & Cham., Linnaea 5: 89. 1830; *Dysosmia acerifolia* M. Roem., Fam. nat. syn. monogr. 2: 151. 1846; *Ceratosepalum micranthum* Oerst., Amér. centr. pl. 17. 1863; *Passiflora ceratosepala* Mast., Trans. Linn. Soc. London. 27: 630. 1871; *Passiflora aspera* Sessé & Moc., Fl. mexic. 227. 1887; *P. scabra* Sessé & Moc., Fl. mexic. 228. 1887.

Description. Vine; stems subterete to subangular, glabrate to hispidulous. Leaves simple, to 13 cm long and 16 cm wide, 3–5-lobed; lobes ovate, glabrate to hispidulous above and below, palmately veined at base; tip acuminate; base cordate; margins entire, denticulate, or serrulate; stipules semiorbicular to reniform, to 1 cm long and 1.5 cm wide; petioles with one conspicuous gland pair near apex. Tendrils simple, axillary. Peduncles solitary or in pairs; bracts 3, to 1 cm long and 0.6 cm wide, lanceolate to oblong, borne at middle of peduncle. Flowers 2–7 cm wide; sepals greenish-white to yellowish, to 4 cm long and 1 cm wide, oblong-lanceolate, corniculate; petals linear-lanceolate, to 1.2 cm long and 0.6 cm wide; corona filaments uniseriate, to 1.8 cm long, filiform, white but purple-banded; ovary subglobose to oblong, densely brownish-tomentose. Fruit greenish-purple to dark purple, to 5 cm long and 3.3 cm wide, ovoid to globose. Seeds obovate to obcordate, to 6 mm long and 4 mm wide, flattened, reticulate.

Phenology. Flowers from July through January. Fruits in May, August, September, November, January, and February.

Distribution. Southern Mexico to Peru. In Costa Rica found from 1,000 to 1,700 m.

Notes. Commonly called *granadilla silvestre* or *comida de culebra.*

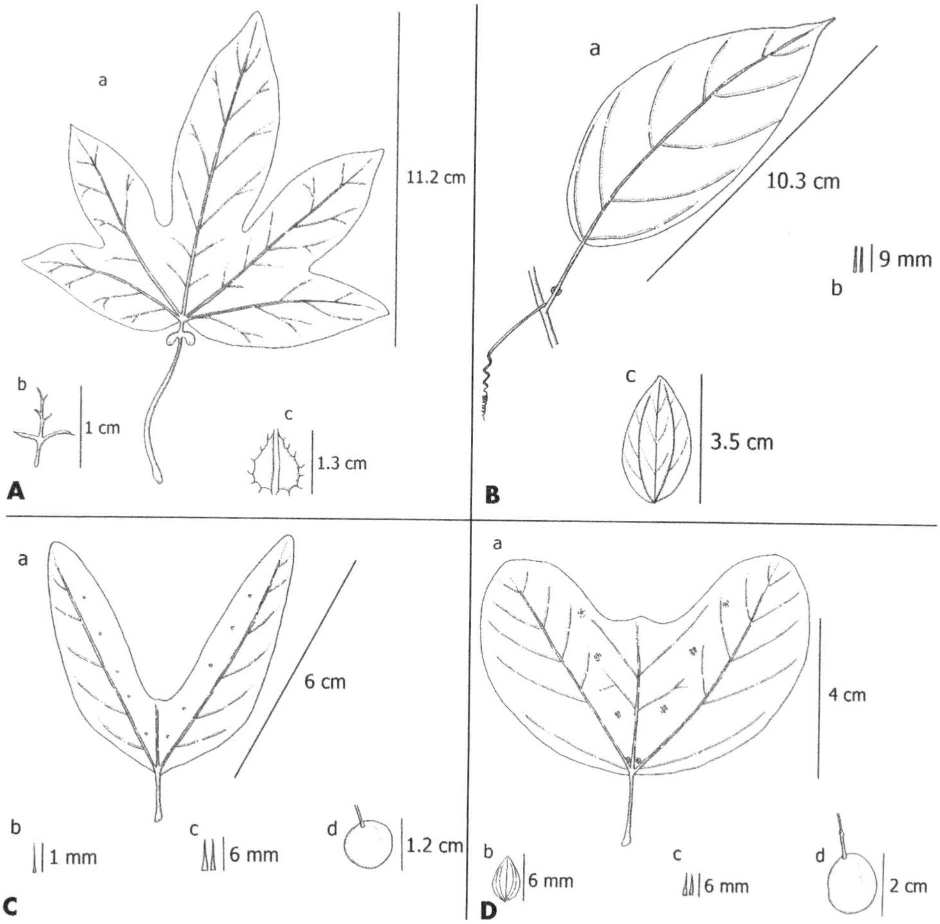

Fig. 7.2. A. *Passiflora adenopoda:* **(a)** leaf, **(b)** floral bract, **(c)** stipules (Leaf and stipules based on Krings 121, CR; Floral bract based on Strother 46909, DUKE). **B.** *Passiflora ambigua:* **(a)** leaf and tendril, **(b)** stipules, **(c)** floral bract (Leaf and tendril based on Haber 11052, INB; Stipules and floral bract based on Manríquez 594, DUKE). **C.** *Passiflora apetala:* **(a)** leaf, **(b)** floral bract, **(c)** stipules, **(d)** fruit (Leaf, stipules, and fruit based on Chavarría 88, CR; Floral bract based on Luteyn 4508, DUKE). **D.** *Passiflora bicornis:* **(a)** leaf, **(b)** floral bract, **(c)** stipules, **(d)** fruit (Leaf, stipules, and fruit based on Ocampo 3228, CR; Floral bract based on Wilbur 26686, DUKE).

2. *Passiflora ambigua* Hemsl., Bot. Mag. 128: pl. 7822. 1902. Figure 7.2B.

Description. Vine or liana; stems terete, glabrous. Leaves simple, to 17 cm long and 8 cm wide, unlobed, oblong-lanceolate to ovate-lanceolate, glabrous above and below, penniveined; tip short-acuminate; base rounded to cuneate; margins entire; stipules filiform, to 9 mm long but sometimes inconspicuous; petioles with one sessile gland pair at or below the middle. Tendrils simple, axillary. Peduncles solitary; bracts ovate, 3–6 cm long, to 4 cm wide, entire, eglandular. Flowers 8–12 cm wide; sepals pinkish and purple puncticulate internally, white externally, to 5 cm long and 1.9 cm wide, more or less linear-lanceolate, short-corniculate beneath apex; petals white with purplish dots, to 4 cm long and 1 cm wide, linear-lanceolate; corona filaments frequently 5-seriate, outermost to 1.5 cm long, red to deep violet with white bands; ovary ovoid, densely brownish-tomentose. Fruit greenish-yellow to speckled bright orange, to 12.5 cm long and 4.5 cm wide, ovoid. Seeds ovate to ovate-oblong, to 7 mm long and 6 mm wide, reticulate.

Phenology. Flowers from January through June and in October. Fruits from April through June and in December.

Distribution. Southern Mexico to Panama. In Costa Rica found from 10 to 900 m.

Notes. Known throughout Central America as *granadilla* or *granadilla de monte* (along with several other species).

3. *Passiflora apetala* Killip, J. Wash. Acad. Sci. 12: 255. 1922. Figure 7.2C.

Description. Vine; stems angulate, sometimes grooved, glabrous. Leaves simple, to 8 cm long and wide, glabrous above and below, glandular-ocellate below, palmately 3-veined from the base, strongly bilobate; lobes often obtuse, mucronate; tip acute; base cuneate to rounded; margins entire; stipules setaceous, to 4 mm long; petioles eglandular. Tendrils simple, axillary. Peduncles paired; bracts setaceous, to 3 mm long. Flowers to 2 cm wide; sepals yellowish-green, to 1 cm long and 0.3 cm wide, more or less oblanceolate, not corniculate; petals lacking or vestigial; corona filaments uni- or biseriate, to 3 mm long, filiform; ovary globose or subglobose, glabrous. Fruit dark bluish to black, to 1 cm long and wide, globose. Seeds ovate, to 3 mm long and 2 mm wide, transversely sulcate.

Phenology. Flowers in October. Fruits from October through June.

Distribution. Costa Rica and Panama. In Costa Rica found from 1,000 to 1,950 m.

4. *Passiflora bicornis* Houston ex Mill., Gard. dict., ed. 8, Passiflora no. 13. 1768. Figure 7.2D.

Synonymy. *Passiflora pulchella* Kunth in H.B.K., Nov. Gen. sp. 2: 134. 1817.

Description. Vine; stems subterete to subangular, glabrous. Leaves simple, to 6 cm long and 10 cm wide, glabrous above and below, ocellate below, bilobate; lobes truncate to rounded at tip; base truncate to subcuneate; margins entire; stipules linear, to 7 mm long; petioles eglandular. Tendrils simple, axillary. Peduncles solitary; bracts often purplish-red, to 1.5 cm long and 1 cm wide, ovate or suborbicular. Flowers to 5.5 cm wide; sepals blue to lavender, to 2 cm long and 0.9 cm wide, oblong; petals blue to lavender, to 1.5 cm long and 0.4 cm wide, ovate- or oblong-lanceolate; corona filaments multiseriate, filiform; inner filaments to 5 mm long; outer filaments to 1.5 cm long; ovary subglobose, glabrous. Fruit globose, to 1.6 cm long and wide. Seeds more or less obovate, to 3 mm long and 1 mm wide, transversely sulcate.

Phenology. Flowers in April, June, and November. Fruits in June, July, and November.

Distribution. Southern Mexico to Colombia and Venezuela. In Costa Rica found from sea level to 800 m.

Notes. Known in El Salvador as *calzoncillo* (as is *P. capsularis*) and *camacarlata*.

5. *Passiflora biflora* Lam., Encycl. 3: 36. 1789. Figure 7.3A.

Synonymy. *Passiflora lunata* var. *costata* Mast. in Mart., Fl. Bras. 13: 552. 1872; *P. brighami* Wats., Proc. Amer. Acad. Arts 21: 473. 1887; *P. transversa* Mast., Bot. Gaz. 16: 7. 1891.

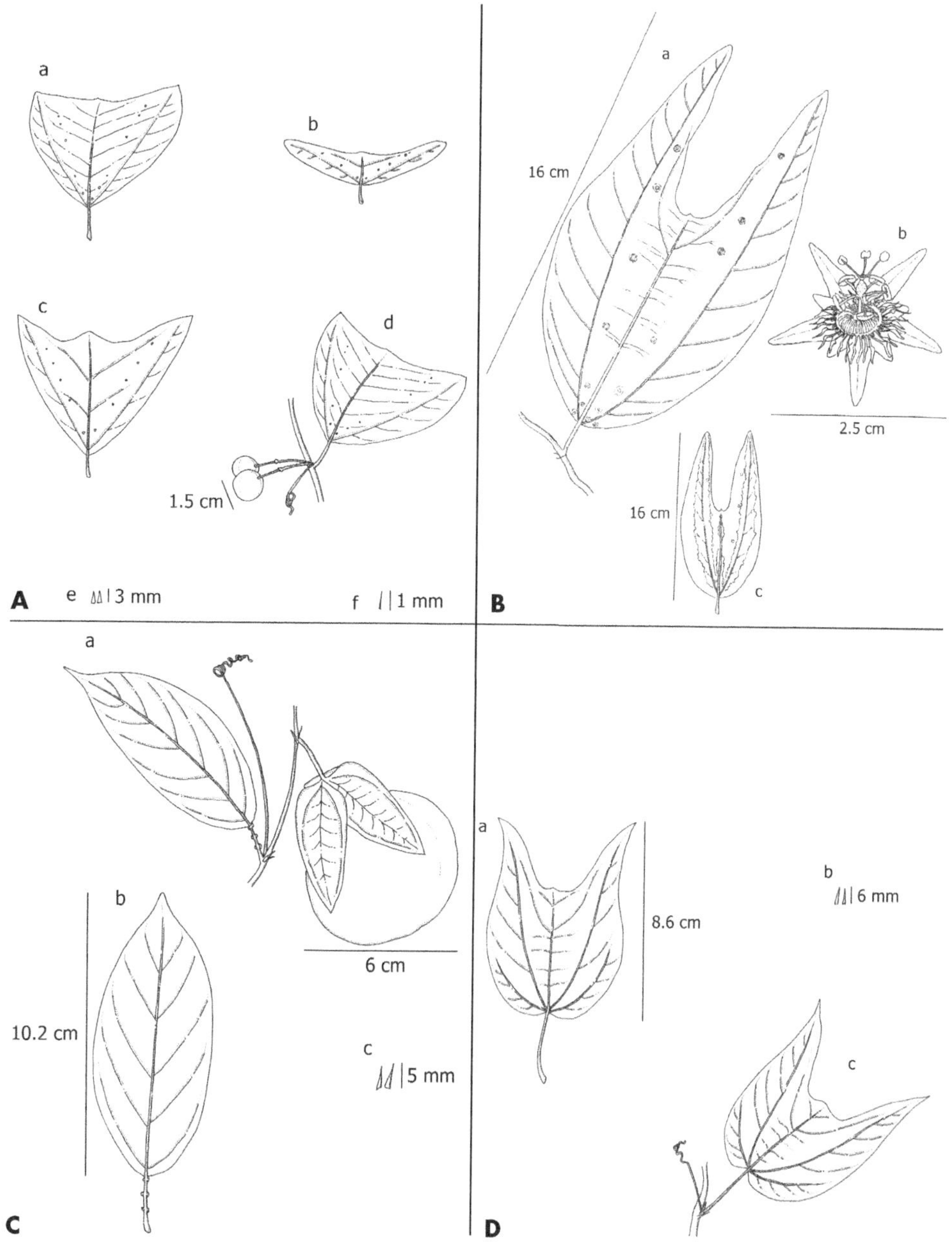

Fig. 7.3. A. *Passiflora biflora:* **(a)** leaf; **(b)** leaf variation; **(c)** leaf variation; **(d)** leaf, tendril, and fruits; **(e)** stipules; **(f)** floral bract (Leaf, stipules, tendril, and fruit based on Martínez 147, INB; Leaf variation based on Morales 4682, INB, and Fuentes 118, INB). **B.** *Passiflora boenderi:* **(a)** leaf, **(b)** flower, **(c)** leaf variegation (Based on Boender BW 92-361, MO). **C.** *Passiflora brevifila:* **(a)** leaf, tendril, and fruit; **(b)** leaf; **(c)** stipules (Based on Krings 23, CR). **D.** *Passiflora capsularis:* **(a)** leaf, **(b)** stipules, **(c)** leaf and tendril (Based on Krings 119, CR).

Description. Vine; stems angulate, sometimes grooved, glabrate. Leaves simple, to 10 cm long and wide (frequently less), glabrous above, glabrous or puberulous and glandular-ocellate below; palmately 3-veined from the base, variable in shape, frequently strongly bilobate; lobes ovate to lanceolate, mucronate; tip acute to rounded; base cuneate, truncate, or rounded; margins entire; stipules setaceous or linear-subulate, to 3 mm long; petioles eglandular. Tendrils simple, axillary. Peduncles paired; bracts setaceous, to 2 mm long. Flowers to 3.5 cm wide; sepals to 1.2 cm long and 0.7 cm wide, puberulent to glabrate and whitish internally, greenish externally, not corniculate; petals white, to 8 mm long and 5 mm wide; corona filaments generally biseriate; inner filaments yellow or yellowish, to 5 mm long, filiform; outer filaments yellow, to 7 mm long, angled and dilated near middle; ovary globose to ovoid, angled or not, glabrate to densely tomentose. Fruit bluish-black, to 2 cm long and wide, globose to ovoid. Seeds obovoid, to 3 mm long and 2 mm wide, transversely sulcate.

Phenology. Flowers and fruits throughout the year.

Distribution. Mexico to Colombia and Venezuela. In Costa Rica found from sea level to 1,700 m.

6. ***Passiflora boenderi*** J.M. MacDougal, Novon 13: 454–457. 2003. Figure 7.3B.

Description. Vine; stems subterete to terete, glabrous. Leaves simple, 11–16 cm long, 5.5–6.5 cm wide, whitish-yellow to yellow variegated, elliptic to ovate, glabrous above and below, ocellate below, palmately 3-veined from the base, strongly bilobate; lobes mucronate; tip acute to acuminate; base rounded; margins entire; stipules narrowly linear to setaceous, 4–5 mm long; petioles eglandular. Tendrils simple, axillary. Peduncles solitary or paired(?). Flowers to 2.5 cm wide; sepals greenish-yellow, to 11 mm long and 5 mm wide, triangular or lanceolate-oblong, not corniculate; petals white, to 6 mm long and 3 mm wide; corona filaments biseriate; outer filaments to 4 mm long, purplish to reddish-green at base, yellowish at apex (drying orange); staminal filaments to 3.2 mm long; ovary subglobose or ellipsoid, to 2.5 mm long, puberulent; styles to 4 mm long. Fruits not seen.

Phenology. Flowering and fruiting times unknown.

Distribution. Endemic to Costa Rica.

Notes. *P. boenderi* J.M. MacDougal is a newly discovered species from Costa Rica collected at 800 m in Alajuela. A striking species, the leaves are conspicuously variegated with large, yellowish ocellae. It resembles *P. gilbertiana* (another Costa Rican endemic), but can be distinguished by its longer lateral lobes and nonmalodorous flowers.

7. ***Passiflora brevifila*** Killip, Contr. U.S. Natl. Herb. 35: 16. 1960. Figure 7.3C.

Description. Vine or liana; stems frequently angulate, glabrous. Leaves simple, to 15 cm long and 6 cm wide, unlobed, oblong to oblong-lanceolate, glabrous above and below, penniveined with arcuately ascending secondary veins; tip acute to acuminate; base rounded to truncate; margins entire; stipules subfalcate to linear-lanceolate, to 6 mm long; petioles with 2–3 gland pairs near apex. Tendrils simple, axillary. Peduncles solitary; bracts lanceolate, 5–7 cm long, to 3 cm wide, entire, acute, and mucronate at apex, subtruncate at base. Flowers to 7 cm wide; sepals whitish, to 2.5 cm long and 0.9 cm wide, oblong, not corniculate; petals whitish, to 2.5 cm long and 0.9 cm wide, oblong; corona filaments uniseriate, to 1 cm long; ovary ovoid. Fruit greenish-yellow to orange, to 7 cm long and wide, globose. Seeds obovate, to 1 cm long and 0.5 cm wide, flattened, reticulate.

Phenology. Flowers in May and October. Fruits from November through January and in June.
Distribution. Costa Rica and Panama. In Costa Rica found from 825 to 1,900 m.

8. *Passiflora capsularis* L., Sp. pl. 957. 1753. Figure 7.3D.

Synonymy. *Granadilla capsularis* (L.) Medik., Malvenfam. 96. 1787; *Passiflora pubescens* Kunth in H.B.K., Nov. Gen. sp. pl. 2: 132. 1817; *Xerogona biloba* Raf., Fl. tellur. 4: 103. 1838; *Passiflora piligera* Gardner, London J. Bot. 1: 173. 1842; *Cieca pubescens* M. Roem., Fam. nat. syn. monogr. 2: 141. 1846; *Decaloba capsularis* M. Roem., Fam. nat. syn. monogr. 2: 154. 1846; *D. smithii* M. Roem., Fam. nat. syn. monogr. 2: 161. 1846; *D. piligera* M. Roem., Fam. nat. syn. monogr. 2: 161. 1846; *Passiflora paraguayensis* Chodat, Bull. Herb. Boissier 7 (app. 1): 74. 1899; *P. hassleriana* Chodat, Bull. Herb. Boissier 7 (app. 1): 74. 1899; *P. hassleriana* var. *grandifolia* Chodat & Hassl., Bull. Herb. Boissier 2: 62. 1904; *P. hassleriana* var. *paraguariensis* Chodat & Hassl., Bull. Herb. Boissier 2: 62. 1904; *P. quinquangularis* Calderón, Passifl. Dilob. 6. Pre-1936.
Description. Vine; stems angulate, glabrous to pubescent. Leaves simple, to 10 cm long and wide, glabrous to puberulous above, densely pubescent below, palmately 3-veined from the base, strongly bilobate; lobes lanceolate, mucronate; tip acute; base cordate to auriculate; margins entire; stipules linear-subulate, to 7 mm long, sometimes falcate; petioles eglandular. Tendrils simple, axillary. Peduncles solitary; bracts absent. Flowers to 6 cm wide; sepals greenish-white to yellowish-green, to 3 cm long and 0.4 cm wide, linear-lanceolate, frequently pilose externally; petals greenish-white to yellowish-green, to 1.5 cm long and 0.4 cm wide, subspatulate, oblong, or lanceolate; corona filaments violet or yellowish-green, to 1.5 cm long, uni- or biseriate; inner filaments capillary, to 3 mm long; outer filaments filiform; ovary ovoid to obovoid, puberulent but sometimes glabrate. Fruit dark purplish, to 6 cm long and 2 cm wide, ellipsoid to fusiform, hexagonal in cross-section. Seeds ovate, to 3+ mm long, transversely sulcate.
Phenology. Flowers in October and November. Fruits in November, December, and June.
Distribution. Guatemala to Brazil and Paraguay, also Greater Antilles. In Costa Rica found from 400 to 1,500 m.
Notes. Known as *calzoncillo* in El Salvador.

9. *Passiflora coriacea* Juss., Ann. Mus. Natl. Hist. Nat. 6: 109, t. 39, f.2. 1805. Figure 7.4A.

Synonymy. *Passiflora difformis* Kunth in H.B.K., Nov. Gen. sp. pl. 2: 136. 1817; *P. clypeata* Sm., Cycl. 26, Passiflora No. 20. 1819; *Monactineirma coriacea* (Juss.) Bory, Ann. Sci. Gén. Phys. Bruxelles 2: 138. 1819; *Cieca difformis* (Kunth) M. Roem., Fam. nat. syn. monogr. 2: 140. 1846; *C. coriacea* (Juss.) M. Roem., Fam. nat. syn. monogr. 2: 148. 1846; *Passiflora sexocellata* Schltdl., Linnaea 27: 521. 1854; *P. obtusifolia* Sessé & Moc., Pl. nov. Hisp. 156. 1887; *P. cheiroptera* Cortés, Fl. Colombia, ed.2, plate between p. 112 and p. 113. 1919, plate only.
Description. Vine; stems angulate, glabrate to puberulent. Leaves simple, to 7 cm long and 25 cm wide, transversely oblong-elliptic, glabrous above and below, glandular ocellate below (ocellae generally restricted to leaf margins), bilobate; lobes divaricate at ninety degrees from midrib; middle lobe usually absent but sometimes strongly developed, mucronate; tip acute to acuminate; base peltate; margins entire; stipules narrowly linear, to 5 mm long; petioles with 1 gland pair at or below middle (rarely with 2 gland pairs or glandu-

lar near apex). Tendrils simple, axillary. Upper inflorescence terminal, to 6 cm long, raceme, leafless, stipules and biglandular bracteiform appendages present; lower inflorescence axillary with single or paired flowers. Flowers to 3.5 cm wide; sepals yellowish-green, to 1.5 cm long and 0.5 cm wide, oblong-lanceolate; petals absent; corona filaments yellow, biseriate; inner filaments to 2 mm long; outer filaments filiform, to 8 mm long; ovary ovoid, glabrous. Fruit bluish, to 2 cm long and wide, globose. Seeds usually obovate, to 4 mm long and 2 mm wide, reticulate, curved and distinctly beaked.

Phenology. Flowers from November through April, and in June and July. Fruits throughout the year.

Distribution. Mexico to Boliva, and possibly British Guiana. In Costa Rica found from sea level to 1,100 m.

Notes. Known variously throughout Central America as *murcielago*, *hoja de murcielago*, *ala de murcielago*, *media luna*, *granadilla de ratón*, and *granadilla del monte*.

10. *Passiflora costaricensis* Killip, J. Wash. Acad. Sci. 12: 257. 1922. Figure 7.4B.

Description. Vine; stems triangulate, frequently with the angles ribbed or slightly winged, hirsute (at least on angles). Leaves simple, to 13 cm long and 11 cm wide, oblong to ovate, conspicuously hirsute below (often dense), palmately 3-veined from the base, strongly bilobate; lobes mucronate; tip acute; base rounded-truncate to cordate; margins entire; stipules subulate, to 8 mm long; petioles eglandular, often densely hirsute. Tendrils simple, axillary. Peduncles solitary; bracts absent. Flowers to 5 cm wide; sepals white, to 2 cm long and 0.4 cm wide, glabrous internally, hirsute externally, not corniculate; petals linear-oblong, to 8 mm long and 2 mm wide; corona filaments generally uniseriate, to 8 mm long; ovary ovoid, minutely puberulent. Fruit reddish, to 8 cm long and 2 cm wide, ellipsoid, asymmetrical in cross-section. Seeds more or less oblong, to 3 mm long and 2 mm wide, somewhat flattened, transversely sulcate.

Phenology. Flowers from April through December. Fruits from May through August, and in November.

Distribution. Guatemala to Costa Rica. In Costa Rica found from 100 to 1,000 m.

11. *Passiflora dioscoreifolia* Killip, J. Wash. Acad. Sci. 14: 108. 1924. Figure 7.4C.

Description. Vine; stems frequently angulate, glabrous to somewhat pubescent (at least at nodes). Leaves simple, to 15 cm long and 8 cm wide, broadly lanceolate to oblong-ovate, glabrous to sparsely pubescent above, glabrous below, sometimes glandular-punctate above and below, palmately 3-veined from base, unlobed; tip acute to acuminate; base cordate or sometimes subcordate; margins entire; stipules broadly ovate to reniform, to 1.6 cm long and 0.8 cm wide; petioles generally with 1 gland pair near apex. Tendrils simple, axillary. Peduncles solitary or paired; bracts setaceous, to 4 mm long, scattered, acute. Flowers to 5 cm wide; sepals to 2 cm long and 1 cm wide, white with purple streaks and glabrous internally, greenish-white and pubescent externally, ovate-lanceolate, not corniculate; petals whitish, to 1.2 cm long and 0.6 cm wide, oblong; corona filaments uniseriate, to 1.5 cm long, white but spotted with purple; ovary ovoid, glabrous. Fruit reddish, to 13+ cm long and 3.5 cm wide, ovoid. Seeds obcordate, to 7 mm long and 5 mm wide, reticulate.

Phenology. Flowers in February, April through June, August, and December. Fruits in November.

Distribution. Endemic to Costa Rica. Found from 1,200 to 1,600 m.

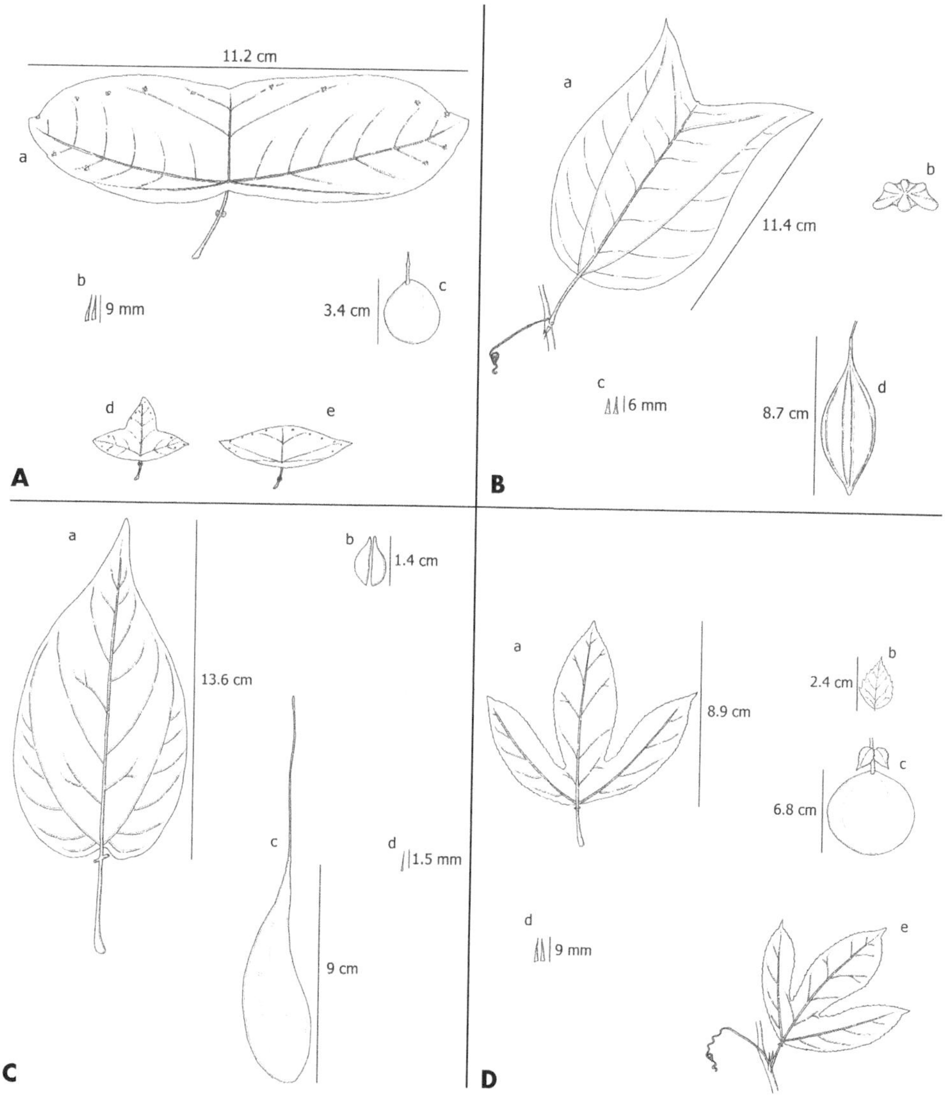

Fig. 7.4. A. *Passiflora coriacea:* **(a)** leaf, **(b)** stipules, **(c)** fruit, **(d & e)** leaf variations (Based on Hammel 17870, CR). **B.** *Passiflora costaricensis:* **(a)** leaf and tendril, **(b)** front view of fruit, **(c)** stipules, **(d)** fruit (Based on Segura 229, INB). **C.** *Passiflora dioscoreifolia:* **(a)** leaf, **(b)** floral bracts, **(c)** fruit, **(d)** stipules (Leaf, stipules, and fruit based on Krings 116, CR; Floral bract based on Taylor 3012, DUKE). **D.** *Passiflora edulis:* **(a)** leaf, **(b)** floral bract, **(c)** fruit, **(d)** stipules, **(e)** leaf and tendril (Leaf, tendril, stipules, and fruit based on Hammel 19798, INB; Floral bract based on Chacón 1102, DUKE).

12. *Passiflora edulis* Sims, Bot. Mag. 45: pl. 1989. 1818. Figure 7.4D.

Synonymy. *Passiflora pallidiflora* Bertol., Syll. Pl. Hort. Bonon. 6. 1827; *P. verrucifera* Lindl., Bot. Reg. 26: pl. 52. 1840; *P. middletoniana* Paxton, Mag. 9: pl. 51. 1842; *P. rigidula* Jacq., Eclog. Pl. 2: pl. 124. 1844 for plate; *P. rubricaulis* Jacq., Eclog. Pl. 2: pl. 169. 1844; *P. pomifera* M. Roem., Fam. nat. syn. monogr. 2: 179. 1846; *P. edulis* var. *verrucifera*

Mast., Trans. Linn. Soc. London 27: 637. 1871; *P. edulis* var. *pomifera* Mast., Trans. Linn. Soc. London 27: 637. 1871; *P. edulis* var. *rubricaulis* Mast., Trans. Linn. Soc. London 27: 637. 1871; *P. vernicosa* Barb. Rodr., Contr. Jard. Bot. Rio de Janeiro 2: 62, pl. 9A. 1902.

Description. Vine; stems terete to subangulate, glabrous to slightly puberulent. Leaves simple, to 11 cm long and 10 cm wide, glabrous above and below, 3-lobate; tip acute to acuminate; base rounded to subcordate; margins serrate; stipules linear-subulate, to 1 cm long; petioles generally with 1 gland pair near apex. Tendrils simple, axillary. Peduncles to 6 cm; bracts ovate, to 2.5 cm long and 1.5 cm wide, serrate to dentate (frequently glandular along margins). Flowers to 7 cm wide; sepals to 3.5 cm long and 1 cm wide, white internally, green externally, oblong; petals white, to 3 cm long, oblong; corona filaments 4–5-seriate; outer two series of filaments white at apex, purple at base, to 2.5 cm long, filiform; ovary ovoid to globose, glabrous to tomentose. Fruit yellowish-greenish to purple, to 5 cm long and wide, globose to ovoid. Seeds more or less ovoid, to 6 mm long and 4 mm wide, minutely reticulate.

Phenology. Flowers from March through May, and in August. Fruits from April through September.

Distribution. South America, but cultivated around the world and naturalized in many areas. In Costa Rica found from 1,500 to 1,800 m.

Notes. The *maracuya* of commerce is a cultivar of this species.

13. *Passiflora eueidipabulum* S. Knapp & Mallet, Ann. Missouri Bot. Gard. 71: 1070. 1984. Figure 7.5A.

Description. Liana, stems terete. Leaves simple, to 18 cm long and 12 cm wide, unlobed, ovate, peltate, glabrous above, papillose below; tip acute; base cordate to rounded; margins entire; stipules setaceous, to 1 mm long; petioles with 2–3 gland pairs (glands sometimes subopposite). Tendrils simple, axillary. Peduncles paired; bracts 3, to 3 mm long, scattered. Flowers borne in pairs on tendrils; to 5 cm wide; sepals white with a greenish stripe externally, to 2.5 cm long and 1.5 cm wide, broadly triangular, not corniculate; petals whitish, to 2.5 cm long and 1.5 cm wide, broadly triangular; corona filaments 3-seriate; outer filaments to 1.5 cm long, mottled green and maroon at base, bright yellow at apex; ovary ellipsoid, to 5 mm long, pubescent. Fruit yellowish-green, to 7 cm long and 5 cm wide, ovoid. Seeds elliptic, to 1 mm long and nearly 4 mm wide, winged.

Phenology. Flowering times unknown. Fruits in Panama in September.

Distribution. Costa Rica and Panama. In Costa Rica found from 250 to 950 m.

14. *Passiflora gilbertiana* J.M. MacDougal, Ann. Missouri Bot. Gard. 76: 610. 1989. Figure 7.5B.

Description. Vine; stems subterete, puberulent. Leaves simple, to 17 cm long and 6.5 cm wide, variegated red and purple, elliptic to obovate, glabrous above, glabrous to puberulent and ocellate below, palmately 3-veined from the base, strongly bilobate; lobes mucronate; tip acute to acuminate; base cuneate to narrowly rounded; margins entire; stipules narrowly lanceolate or linear, to 6 mm long; petioles eglandular, puberulent. Tendrils simple, axillary. Peduncles paired; bracts 3, to 5 mm long, linear-triangular. Flowers to 2 cm wide; sepals greenish-yellow, to 1.5 cm long and 0.6 cm wide, triangular to lanceolate-oblong, somewhat puberulous, not corniculate; petals greenish-yellow, to nearly 4 mm long and

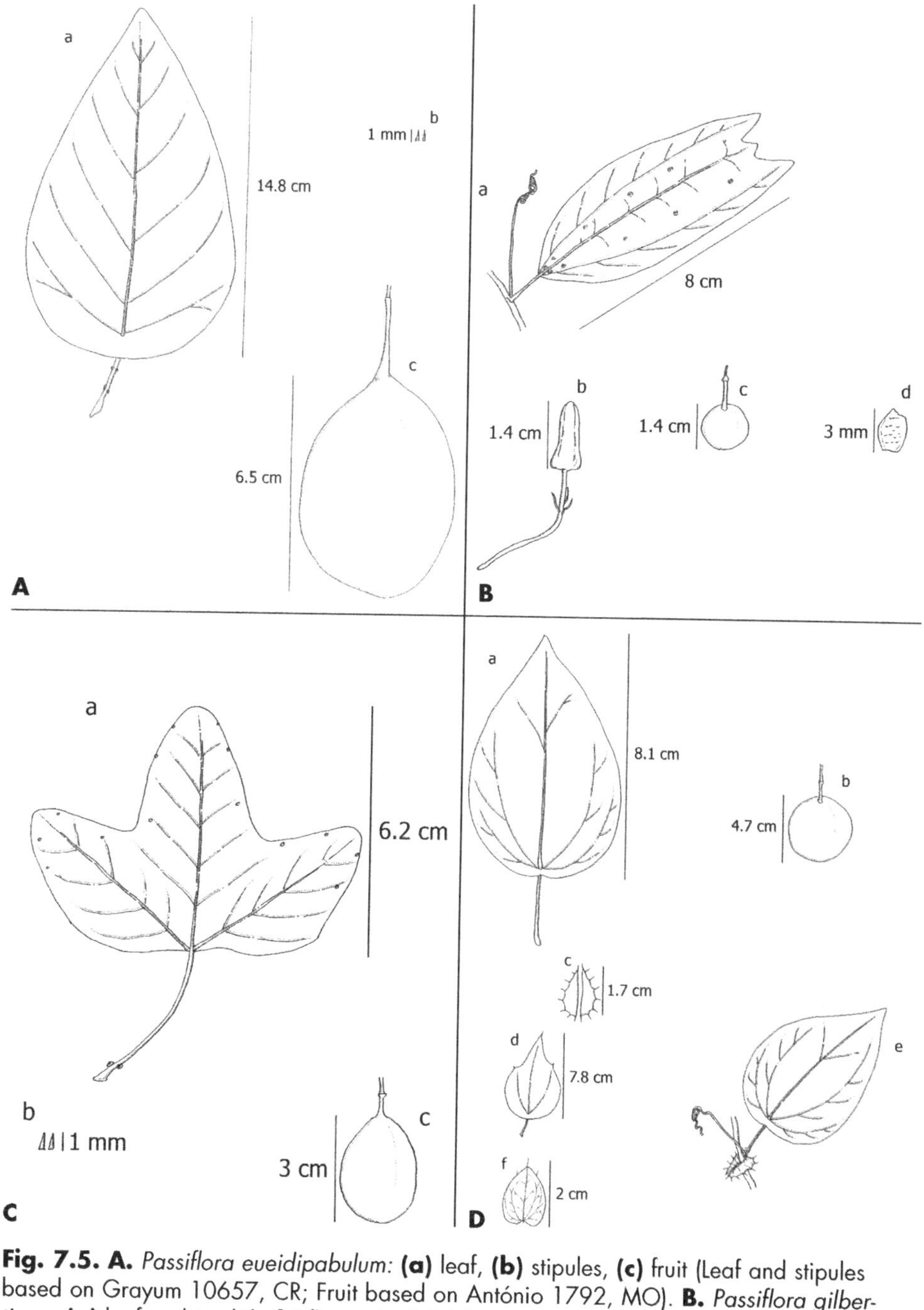

Fig. 7.5. A. *Passiflora eueidipabulum:* **(a)** leaf, **(b)** stipules, **(c)** fruit (Leaf and stipules based on Grayum 10657, CR; Fruit based on António 1792, MO). **B.** *Passiflora gilbertiana:* **(a)** leaf and tendril, **(b)** flower bud, **(c)** fruit, **(d)** seed (Leaf, tendril, fruit, and seed based on Gilbert 8, MO; Flower bud based on Taylor and Skotak 4771, MO). **C.** *Passiflora gracilis:* **(a)** leaf, **(b)** stipules, **(c)** fruit (Based on Solís R. 31 CR). **D.** *Passiflora guatemalensis:* **(a)** leaf; **(b)** fruit; **(c)** stipules; **(d)** leaf variation; **(e)** leaf, tendril, and stipules; **(f)** floral bract (Leaf, leaf variation, tendril, stipules, and fruit based on Chávez 319, INB; Floral bract based on Martínez S. 7008, DUKE).

2 mm wide, oblong or oblong-ovate; corona golden; filaments uni- or biseriate, to 7 mm long; ovary subglobose or ellipsoid, frequently puberulent. Fruit purplish to black, to 1.5 cm long and wide, subglobose to ellipsoid. Seeds obovate, to 3+ mm long and 2 mm wide, transversely sulcate.

Phenology. Flowers in March. Fruits in August.

Distribution. Endemic to Costa Rica. Found at about 2,300 m.

Notes. The scent of the flowers of this species has been variously described as putrid, foul, stinking, and even odor of pigpen.

15. *Passiflora gracilis* Jacq. ex Link, Enum. hort. berol. alt. 2: 182. 1822. Figure 7.5C.

Synonymy. *Cieca gracilis* (Jacq. ex Link) M. Roem., Fam. nat. syn. monogr. 2: 141. 1846.

Description. Vine; stems angulate. Leaves simple, to 7 cm long and 10 cm wide, trilobate with the middle lobe dominant or equal to the lateral lobes, glabrous above and below, glandular-ocellate below near margins, palmately 3-veined from the base; tip obtuse to rounded; base cordate; margins entire; stipules narrowly linear, to 7 mm long; petioles with 1 gland pair (sometimes subopposite) in the lower half. Tendrils simple, axillary. Peduncles to 3 cm long; bracts setaceous, to 2 mm long. Flowers to 2 cm wide; sepals white, to 1 cm long and 0.3 cm wide, narrowly oblong, not corniculate; petals absent; corona filaments white and mauve at base, generally biseriate; inner filaments capillary, slightly longer than 1 mm; outer filaments filiform, to 8 mm long; ovary ovoid, glaucous. Fruit purplish to scarlet, to 2.5 cm long and 1.5 cm wide, ellipsoid. Seeds subglobose, to 4 mm long and 3 mm wide, flattened, reticulate.

Phenology. Flowers in May and September. Fruits in December and January.

Distribution. Found in Costa Rica to 1,100+ m.

16. *Passiflora guatemalensis* S. Watson, Proc. Amer. Acad. Arts 22: 473. 1887. Figure 7.5D.

Description. Vine; stems terete to angulate, glabrous. Leaves simple, to 8 cm long and 7 cm wide, ovate-orbicular to orbicular, glabrous above and below, palmately 3–5-veined from the base, peltate, sometimes shallowly trilobate at apex; lobes mucronate; tip acute; base rounded to subcordate; margins entire; stipules reniform, to 1 cm long and 2 cm wide, setiferous-dentate; petioles eglandular. Tendrils simple, axillary. Peduncles solitary; bracts cordate or rounded-cordate, to 3 cm long and 2 cm wide, denticulate and somewhat setaceous. Flowers to 6 cm wide; sepals white to creamish, to 3 cm long and 1 cm wide, oblong, not corniculate; petals white to creamish, to 3 cm long and 1 cm wide, oblong; corona filaments yellow, biseriate; inner filaments less than 5 mm; outer filaments to 1.5 cm long; ovary ovoid, glabrous. Fruit purple, to 3.5 cm long and wide, globose. Seeds oblong-ovate, to 7 mm long and 4 mm wide, flattened, reticulate.

Phenology. Flowering times unknown. Fruits in February, April, June, September, and October.

Distribution. Southern Mexico to Colombia. In Costa Rica found from 800 to 1,100 m.

17. *Passiflora hahnii* (Fourn.) Mast., Trans. Linn. Soc. London. 27: 638. 1871. Figure 7.6A.

Synonymy. *Disemma hahnii* Fourn., Rev. hort. 40: 430, pl. opp. 430. 1869; *Passiflora cookii* Killip, J. Wash. Acad. Sci. 12: 256. 1922.

Description. Vine; stems glabrous. Leaves simple, to 8 cm long and 7 cm wide, suborbicular to orbicular, peltate, glabrous above and below, palmately 3–5-veined from the base, sometimes shallowly trilobate at apex; lobes mucronate; tip acute; base rounded to subcordate; margins entire, sometimes with glands near the base; stipules reniform, to 1.5 cm long, 3 cm wide, cordate, minutely crenulate; petioles eglandular. Tendrils simple, axillary. Peduncles solitary, 2–7.5 cm long; bracts pale green to reddish-green, to 6 cm long and 4 cm wide, cordate-ovate or -reniform, rounded or obtuse at apex. Flowers to 4.5 cm wide; sepals ovate-lanceolate, to 1.5 cm long and 1.2 cm wide, not corniculate; petals white, to 1.5 cm long and 0.8 cm wide, ovate-lanceolate; corona filaments biseriate; inner filaments nearly 3 mm long; outer filaments to 1 cm long; ovary subglobose. Fruit ellipsoid. Seeds flattened, to 5 mm long. Reticulate.

Phenology. Flowering times unknown. Fruits in March.

Distribution. Mexico to Costa Rica. In Costa Rica found from sea level to 1,850 m.

Notes. The name *P. hahnii* (Fourn.) Mast. was misapplied by Killip (1938) to the species here called *P. guatemalensis* S. Watson. MacDougal (MO) in a 1990 annotation on a specimen at US noted that the type of *P. hahnii* belongs to what Killip called *P. cookii*.

Passiflora hahnii is related to *P. guatemalensis* and *P. membranacea*, but is distinguished from the former by its minutely crenate stipules, and from the latter by its shorter peduncles (2–7.5 cm). Although peduncles of *P. membranacea* can be as short as 6 cm, all plants have at least some peduncles that are longer than 7.5 cm.

18. *Passiflora helleri* Peyr., Linnaea 30: 54. 1859. Figure 7.6B.

Synonymy. *Passiflora fuscinata* Mast. in Mart., Fl. Bras. 13: 551. 1872.

Description. Vine; stems glabrate to pubescent. Leaves simple, to 8 cm long and 7 cm wide, ovate-oblong to suborbicular, glabrous to minutely puberulous and ocellate below, palmately 3-veined from the base, frequently shallowly trilobate with the middle lobe dominant; lobes mucronate; tip acute; base rounded to subcuneate; margins entire; stipules linear-subulate, to 3 mm long; petioles eglandular. Tendrils simple, axillary. Peduncles frequently paired, to 3.5 cm; bracts setaceous, to 3 mm long. Flowers to 4 cm wide; sepals to 1.5 cm long, whitish internally, green externally, oblong-lanceolate, not corniculate; petals white or pinkish, to 1 cm long and 0.4 cm wide, narrowly oblong; corona filaments greenish with purple dots, to 7 mm long, uniseriate; ovary ovoid, minutely pilose in dense to scattered patches. Fruit dark-green to black, to 2.5 cm long and wide, globose, glabrate. Seeds ovoid, to 4 mm long and nearly 3 mm wide, somewhat flattened, transversely sulcate.

Phenology. Flowering times unknown. Fruits in June and July.

Distribution. Mexico to Costa Rica. In Costa Rica found from 1,300 to 1,520 m.

Notes. *Passiflora helleri* is similar to *P. talamancensis* in general appearance, although its leaves are at most only 1.5 times longer than wide, and never 2 times longer than wide as is the case with *P. talamancensis*. The staminal filaments are also longer in *P. helleri*, often to 1 cm or more, whereas they are frequently shorter than 0.7 cm in *P. talamancensis*.

19. *Passiflora holosericea* L., Sp. pl. 958. 1753. Figure 7.6C.

Synonymy. *Decaloba holosericea* (L.) M. Roem., Fam. nat. syn. monogr. 2: 164. 1846; *Passiflora reticulata* C. Wright in Sauv., Fl. cub. 56. 1873; *P. tuxtlensis* Sessé & Moc., Fl. mexic. 229. 1887.

Description. Vine; stems terete, velutinous to tomentose. Leaves simple, to 14 cm long and 13

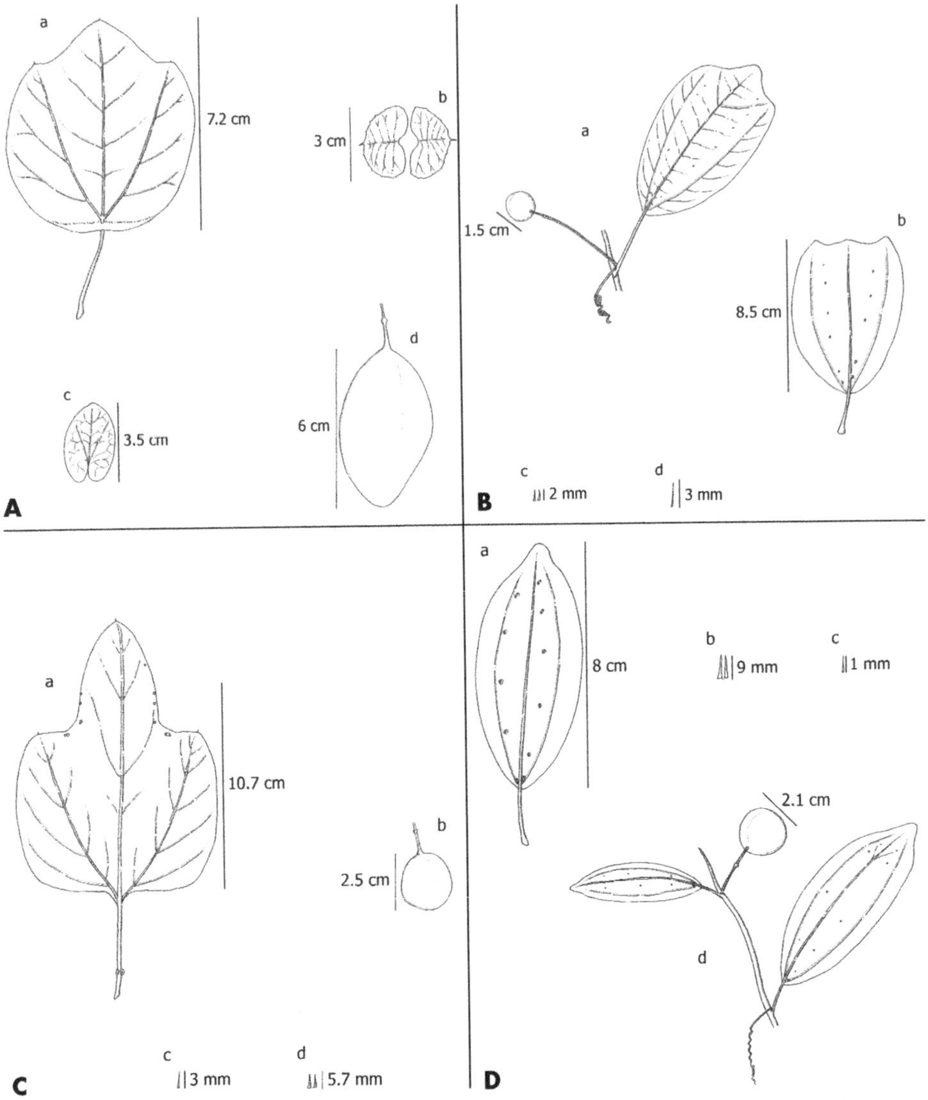

Fig. 7.6. A. *Passiflora hahnii:* **(a)** leaf, **(b)** stipules, **(c)** floral bract, **(d)** fruit (Leaf, stipules, and fruit based on Davidse et al. 25663, MO; Floral bract based on MacDougal and Miley 556, DUKE). **B.** *Passiflora helleri:* **(a)** leaf, tendril, and fruit; **(b)** leaf variation; **(c)** stipules; **(d)** floral bract (Leaf, tendril, and fruit based on Haber 9165, INB; Leaf variation based on Türckheim 1618 155, US; Floral bract based on Perino 3196, DUKE). **C.** *Passiflora holosericea:* **(a)** leaf, **(b)** fruit, **(c)** floral bract, **(d)** stipules (Leaf, stipules, and fruit based on Rivera 535, CR; Floral bract based on MacDougal and Miley 491, DUKE). **D.** *Passiflora lancearia:* **(a)** leaf; **(b)** stipules; **(c)** floral bract; **(d)** leaves, tendril, and fruit (Leaves, stipules, tendril, and fruit based on Morales 4078, INB; Floral bract based on Hammel et al. 10918, DUKE).

cm wide, velutinous above, tomentose and ocellate below (ocellae usually restricted to margins of lobes), palmately 3-veined from the base, trilobate with the middle lobe dominant; lobes mucronate; tip rounded to acute; base cordate; margins entire; stipules filiform, to 6 mm long; petioles with 1 gland pair near middle. Tendrils simple, axillary. Peduncles solitary or paired; bracts linear-subulate, to 3 mm long. Flowers to 4 cm wide; sepals white, to 1.5 cm long, glabrous and spotted with red (sometimes sparsely) internally, densely pubescent externally; petals white with brownish streaks or reddish mottles, to 1.3 cm long and 0.6 cm wide, oblanceolate to spatulate; corona filaments generally biseriate; inner filaments capillary, to 5 mm long; outer filaments yellow at apex and purple at base, to 7 mm long, lanceolate; ovary obovoid, frequently pilose. Fruit yellowish-green, to 2.5 cm long and wide, globose. Seeds more or less obovate, to 3 mm long and 2 mm wide, reticulate.

Phenology. Flowering times unknown. Fruits from July through October, and in March.

Distribution. Mexico to northern South America. In Costa Rica found from sea level to 1,000+ m.

20. *Passiflora lancearia* Mast., J. Bot. 23: 114. 1885. Figure 7.6D.

Description. Vine; stems glabrate to puberulous. Leaves simple, to 8 cm long and 4 cm wide; oblong to oblong-lanceolate, glabrous above and below, ocellate below, palmately 3-veined from the base, frequently shallowly lobate with the middle lobe dominant; lobes mucronate; tip rounded to acuminate; base rounded to cuneate; margins entire; stipules setaceous, to 1 cm long; petioles eglandular. Tendrils simple, axillary. Peduncles paired; bracts setaceous, 1 mm long. Flowers to 4 cm wide; sepals whitish, to 1.5 cm long and 1 cm wide, tomentose internally, green externally, ovate-lanceolate, not corniculate; petals white, to 8 mm long and 5 mm wide; corona filaments generally biseriate; inner filaments capillary, 2 mm long; outer filaments filiform, to 5 mm long; ovary ovoid, glabrous to short-pubescent. Fruit green, to 4.6 cm long and 3.5 cm wide, subglobose to ellipsoid. Seeds ovate, to 6 mm long and nearly 5 mm wide, flattened, transversely sulcate.

Phenology. Flowers from February through June. Fruits from October through January, and from March through July.

Distribution. Costa Rica and Panama. In Costa Rica found from sea level to 1,800 m.

21. *Passiflora ligularis* Juss., Ann. Mus. Natl. Hist. Nat. 6: 113, t. 40. 1805. Figure 7.7A.

Synonymy. *Passiflora ligularis* var. *geminiflora* DC., Prodr. 3: 328. 1828; *P. serratistipula* DC., Prodr. 3: 328. 1828.

Description. Vine; stems terete, glabrous. Leaves simple, to 18 cm long and 14 cm wide, broadly lanceolate to broadly ovate, glabrous above and below, unlobed; tip acute to acuminate; base cordate; margins entire; stipules ovate-lanceolate to oblong-lanceolate, to 2.5 cm long; petioles with 2–4 scattered, filiform or liguliform gland pairs (glands sometimes subopposite). Tendrils simple, axillary. Peduncles solitary or paired; bracts ovate or ovate-lanceolate, to 4 cm long and 1.5 cm wide, acute at apex. Flowers to 9 cm wide; sepals white internally, green externally, to 3.5 cm long, ovate-oblong, not corniculate; petals whitish or pinkish, to 3 cm long and 1 cm wide, more or less oblong; corona filaments 5–7-seriate; inner filaments nearly 2 mm long; outer filaments blue at apex, white and purplish banded elsewhere, to 3 cm long; ovary ovoid. Fruit yellowish, purplish, or reddish, to 8 cm long and 5 cm wide, ovoid to ellipsoid. Seeds obcordate, to 6 mm long and 4 mm wide, reticulate; apex sometimes minutely 3-dentate.

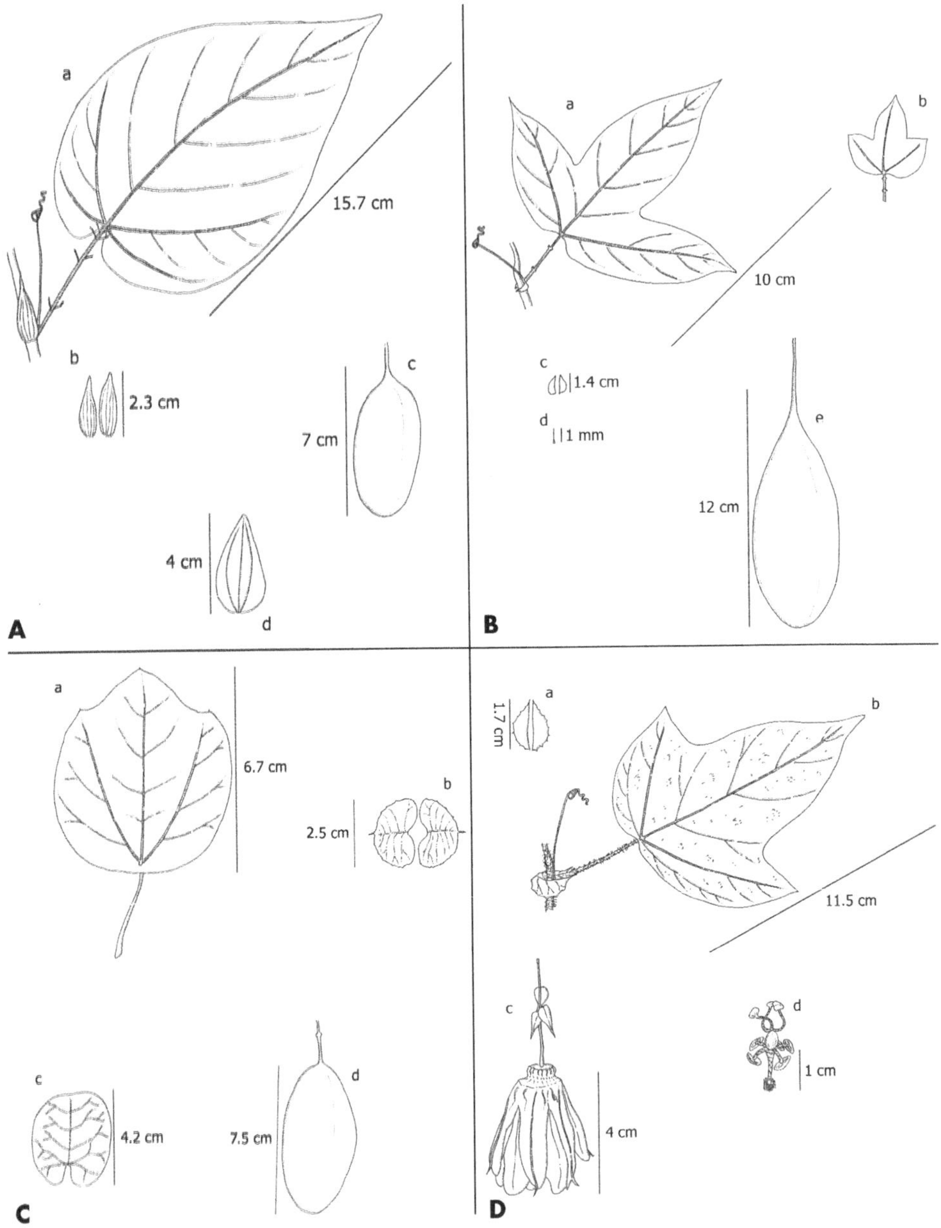

Fig. 7.7. A. *Passiflora ligularis:* **(a)** leaf, tendril, and stipules; **(b)** stipules; **(c)** fruit; **(d)** floral bract (Leaf, tendril, stipules, and fruit based on Morales 1325, INB; Floral bract based on Taylor and Skotak 4527, DUKE). **B.** *Passiflora lobata:* **(a)** leaf, stipules, and tendril; **(b)** leaf variation; **(c)** stipules; **(d)** floral bract; **(e)** fruit (Leaves, stipules, tendril, and fruit based on Kernan 387, INB; Floral bract based on Hammel 8007, DUKE). **C.** *Passiflora membranacea:* **(a)** leaf, **(b)** stipules, **(c)** floral bract, **(d)** fruit (Leaf, stipules, and fruit based on Hatheway 1386, CR; Floral bract based on Webster et al. 12231, DUKE). **D.** *Passiflora menispermifolia:* **(a)** stipules; **(b)** leaf, stipules, and tendril; **(c)** unopened flower showing corniculate sepals; **(d)** androgynophore (Based on Krings 150, CR).

Phenology. Flowers in September. Fruits in October.
Distribution. Central Mexico to South America. In Costa Rica found from 1,000 to 1,950+ m.
Notes. Grown in Central America for its tasty fruits. This species is the *granadilla* most commonly found in markets.

22. *Passiflora lobata* (Killip) Hutch. ex J.M. MacDougal, Phytologia 60: 446. 1986. Figure 7.7B.

Synonymy. *Tetrastylis lobata* Killip, J. Wash. Acad. Sci. 16: 368. 1926.
Description. Vine; stems angulate, glabrous. Leaves simple, to 15 cm long and 19 cm wide, slightly hispidulous above, glabrous below, 3-lobate; lobes oblong to oblong-lanceolate; tip acute to acuminate; base cordate; margins entire; stipules semiovate to reniform, to 1.5 cm long and 0.8 cm wide; petioles with 2 gland pairs usually near or below middle. Tendrils simple, axillary. Peduncles solitary or paired; bracts setaceous, to 3 mm long. Flowers to 6 cm wide; sepals whitish streaked with violet internally, green externally, to 2.5 cm long and 0.8 cm wide, oblong to oblong-lanceolate; petals violet-streaked, to 1.5 cm long and 0.7 cm wide, ovate-lanceolate; corona filaments generally uniseriate, to 2 cm long, filiform; ovary ovoid, glabrous. Fruit greenish but sometimes white-spotted, to 9 cm long and 3+ cm wide, obovoid. Seeds obovate, to 9 mm long and 5 mm wide, reticulate.
Phenology. Flowers from July through April. Fruits from February through July, and September through October.
Distribution. Costa Rica to Colombia. In Costa Rica found from sea level to 1,550+ m.

23. *Passiflora membranacea* Benth., Pl. hartw. 83. 1841. Figure 7.7C.

Synonymy. *Cieca membranacea* (Benth.) M. Roem., Fam. nat. syn. monogr. 2: 140. 1846.
Description. Vine; stems glabrous. Leaves simple, to 8 cm long and 7 cm wide, orbicular to suborbicular, peltate, glabrous above and below, palmately 3-veined from the base, sometimes shallowly trilobate at apex; lobes mucronate; tip acute; base rounded to subcordate; margins entire; stipules reniform, to 1.5 cm long and 3 cm wide, cordate, minutely crenulate; petioles eglandular. Tendrils simple, axillary. Peduncles solitary, 6–14 cm long; bracts pale green, red, reddish-green, or reddish-purple, to 5 cm long and 4 cm wide, cordate-ovate, reniform, rounded, or obtuse at apex. Flowers to 8 cm wide; sepals oblong-lanceolate, to 4 cm long and 1 cm wide, not corniculate; petals oblong-elliptic to oblanceolate, to 4 cm long and 1 cm wide; corona filaments biseriate; inner filaments capillary, nearly 2 mm long; outer filaments red, to 1 cm long, filiform; ovary ellipsoid, trisulcate. Fruit green to yellow, to 8 cm long, oblong to ovoid. Seeds oblong-ovate, to 5 mm long and 3 mm wide, punctate.
Phenology. Flowers from February through May, and from September through December. Fruits from July through October, and from January through March.
Distribution. Southern Mexico to Panama. In Costa Rica found from 1,600 to 2,600 m.

24. *Passiflora menispermifolia* Kunth in H.B.K., Nov. Gen. sp. 2: 137. 1817. Figure 7.7D.

Synonymy. *Passiflora cuellensis* Goudot ex Triana & Planch., Ann. Sci. Nat., Bot. Ser. 5: 154. 1873; *P. villosa* Dombey ex Triana & Planch., Ann. Sci. Nat., Bot. Ser. 5: 154. 1873.
Description. Vine; stems glabrate to pilose. Leaves simple, to 14 cm long and 12 cm wide, peltate or basifixed, pubescent or pilose above and below, broadly trilobate; lobes acute

to acuminate at tip; base cordate; margins entire (sometimes inconspicuously serrulate near leaf base); stipules ovate to reniform, to 3.5 cm long, glandular denticulate to subentire; petioles with 2–4 sometimes subopposite gland pairs. Tendrils simple, axillary. Peduncles solitary, to 6 cm long; bracts narrowly lanceolate, to 2 cm long. Flowers to 6 cm wide; sepals greenish, to 4 cm long, oblong, corniculate; petals light purplish, to 4 cm long, linear-oblong; corona filaments purple and white striped or dotted, generally multiseriate; outer filaments to 2 cm long; ovary ovoid. Fruit ovoid, to 6 cm long. Seeds flattened, to 5 mm long and 3 mm wide, reticulate.

Phenology. Flowers from July through December, and in March. Fruits in February, October, and December.

Distribution. Nicaragua to Bolivia. In Costa Rica found from sea level to 1,200 m.

25. *Passiflora mollissima* (Kunth) Bailey, Rhodora 18: 156. 1916. Figure 7.8A.

Synonymy. *Tacsonia mollissima* Kunth in H.B.K., Nov. Gen. sp. pl. 2: 144. 1817.

Description. Vine; stems terete, yellowish-golden villous. Leaves simple, to 10 cm long and 11 cm wide, pubescent to yellowish villous above and below, trilobate; lobes acute at tips; base subcordate; margins serrate; stipules subreniform to reniform, to 1 cm long, subentire to denticulate; petioles with 4–6 sometimes subopposite gland pairs. Tendrils simple, axillary. Peduncles to 6 cm; bracts united half their length or less, to 3.5 cm long, pubescent to tomentose. Flowers to 11 cm wide; calyx tube whitish internally, green to reddish-green externally, to 8+ cm long and 1 cm wide, usually glabrous but sometimes pubescent; sepals oblong, to 5 cm long and 2 cm wide, aristulate or corniculate; petals pink, to 5 cm long and 2 cm wide; corona reduced to a purple band of tubercles or crenulations; ovary oblong, tomentose. Fruit yellowish, to 7 cm long and 3+ cm wide, oblong or ovoid. Seeds obovate, to 6 mm long and 4 mm wide, reticulate.

Phenology. Flowers in March, June, and July. Fruits in March and June.

Distribution. Mexico, Costa Rica, and Panama to Bolivia. In Costa Rica found from sea level to nearly 3,000 m. Also cultivated around the world for the edible fruit, especially in California and Europe.

26. *Passiflora nelsonii* Mast. & Rose, Contr. U.S. Natl. Herb. 5: 142, t. 17. 1897. Figure 7.8B.

Description. Vine or liana; stems terete, glabrous. Leaves simple, to 17 cm long and 10 cm wide, unlobed, broadly ovate, glabrous above and below, penni-veined; tip acuminate; base rounded or cordate; margins entire; stipules oblanceolate, 10–12 mm long; petioles with two gland pairs in upper half of petiole; petiole glands to 1 mm long and 3 mm wide. Tendrils simple, axillary. Peduncles solitary, to 3 cm long; bracts ovate, to 6 cm long and 4.5 cm wide. Flowers 5–7 cm wide; sepals more or less oblong-lanceolate, to 3 cm long, corniculate; petals to 3 cm long; corona filaments multiseriate; outer filaments to 1.5 cm long; ovary obovoid. Fruit subglobose to ellipsoid, to 9 cm long and 6 cm wide. Seeds frequently notched at the base, to 9 mm long.

Phenology. Flowering times unknown. Fruits in May in Mexico. Fruits in Costa Rica in July.

Distribution. Mexico to Panama. In Costa Rica found around 1,100 m.

27. *Passiflora nubicola* J.M. MacDougal, Ann. Missouri Bot. Gard. 76: 608. 1989. Figure 7.8C.

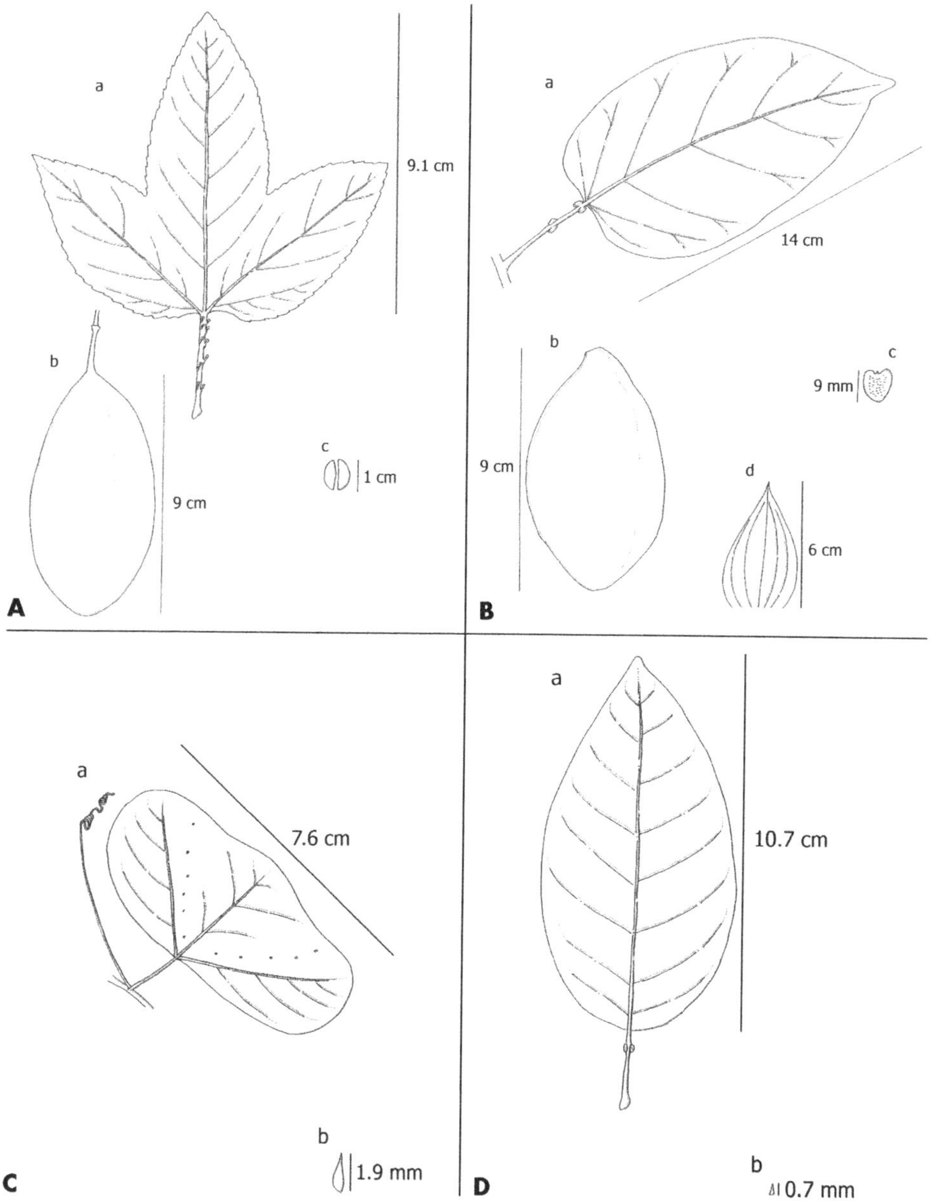

Fig. 7.8. A. *Passiflora mollissima:* **(a)** leaf, **(b)** fruit, **(c)** stipules (Based on Bittner 1065, CR). **B.** *Passiflora nelsonii:* **(a)** leaf, **(b)** fruit, **(c)** seed, **(d)** floral bract (Leaf, floral bract, fruit, and seed based on Gentry et al. 32435, MO). **C.** *Passiflora nubicola:* **(a)** leaf and tendril, **(b)** floral bract (Based on MacDougal 1244, DUKE). **D.** *Passiflora obovata:* **(a)** leaf, **(b)** floral bract (Leaf based on Haber 11052, CR; Floral bract based on Knapp and Mallet 871, DUKE).

Description. Vine; stems angulate, glabrate to puberulent. Leaves simple, to 4.7 cm long and 9 cm wide, obovate to suborbicular, glabrous above, glandular-ocellate and sometimes puberulous below (often restricted to larger veins), palmately 3-veined from the base, shallowly trilobate at apex; lobes mucronate, obtuse, rounded, or truncate at tips; base rounded to subtruncate; margins entire; stipules narrowly linear-triangular, to 2 mm long, inconspicuous; petioles eglandular. Tendrils simple, axillary. Peduncles paired; bracts linear-triangular to narrowly lanceolate, to 2 mm long. Flowers to 2 cm wide; sepals white internally, yellowish-green externally, to 1.1 cm long and 0.4 cm wide, oblong-lanceolate, not corniculate; petals white, to 7 mm long and 3 mm wide, more or less oblong; corona filaments uni- or biseriate; inner filaments capillary when present, 2 mm long; outer filaments yellowish-green to yellow at apex, to 3 mm long; ovary ellipsoid to obovoid, to 2 mm long, glabrous. Fruit green to yellow, to 2.5 cm long and 2.3 cm wide, ellipsoid to obovoid. Seeds obovate, to 5 mm long and 3 mm wide, transversely sulcate.

Phenology. Flowers in April and June. Fruits in January.

Distribution. Endemic to Costa Rica. Found from 1,300 to 1,600 m.

28. *Passiflora obovata* Killip, Publ. Carnegie Inst. Wash. 461: 308, pl.1. 1936. Figure 7.8D.

Description. Vine or liana; stems angulate, glabrous. Leaves simple, to 12 cm long and 6 cm wide, unlobed, obovate to oblong-obovate, glabrous above and below; tip acuminate; base narrowed; margins entire; stipules inconspicuous, setaceous or linear-subulate, deciduous; petioles with one gland pair in upper half. Tendrils simple, axillary. Peduncles paired; bracts inconspicuous, to 7 mm long, triangular-ovate. Flowers to 4 cm wide; sepals more or less oblong, to 1.5 cm long and 0.8 cm wide, not corniculate; petals linear-oblong, to 1.3 cm long and 0.4 cm wide; corona filaments frequently biseriate; inner filaments capillary, to 2 mm long; outer filaments to 1.3 cm long; ovary globose. Fruit globose, to 7 cm wide. Seeds suborbicular, to 8 mm long, reticulate, flattened.

Phenology. Flowers in March, April, and August. Fruits in March, April, August, and October.

Distribution. Mexico to Costa Rica. In Costa Rica found from 500 to 1,400 m.

29. *Passiflora oerstedii* Mast., Fl. Bras. 13, pt. 1: 562. 1872. Figure 7.9A.

Synonymy. *Passiflora populifolia* Triana & Planch., Ann. Sci. Nat., Bot. Ser. 5: 150. 1873; *P. purpusii* Killip, J. Wash. Acad. Sci. 12: 261. 1922; *P. dispar* Killip, J. Wash. Acad. Sci. 12: 330. 1922.

Description. Vine; stems terete to subangulate, glabrous. Leaves simple, to 12 cm long and 8 cm wide, unlobed, ovate-lanceolate to oblong-lanceolate, glabrous above, somewhat pubescent or slightly villous below; tip acute to acuminate; base cordate or subcordate; margins entire; stipules semiovate, to 4 cm long and 1.7 cm wide; petioles generally with 2–3 gland pairs (sometimes subopposite). Tendrils simple, axillary. Peduncles frequently solitary; bracts ovate-lanceolate, to 1.5 cm long, base rounded to cordate. Flowers to 6 cm wide; sepals ovate-lanceolate, to 3 cm long, white internally, corniculate; petals linear to ovate-lanceolate, to 1.5 cm long and about 0.5 cm wide, white to pinkish; corona filaments purple, frequently multiseriate; inner filaments to 3 mm long; outer filaments to 2 cm long; ovary ovoid, glabrous. Fruit ovoid, to 6 cm long and 3 cm wide. Seeds obovate, to 5 mm long and 3 mm wide, reticulate, sometimes shallowly tridentate.

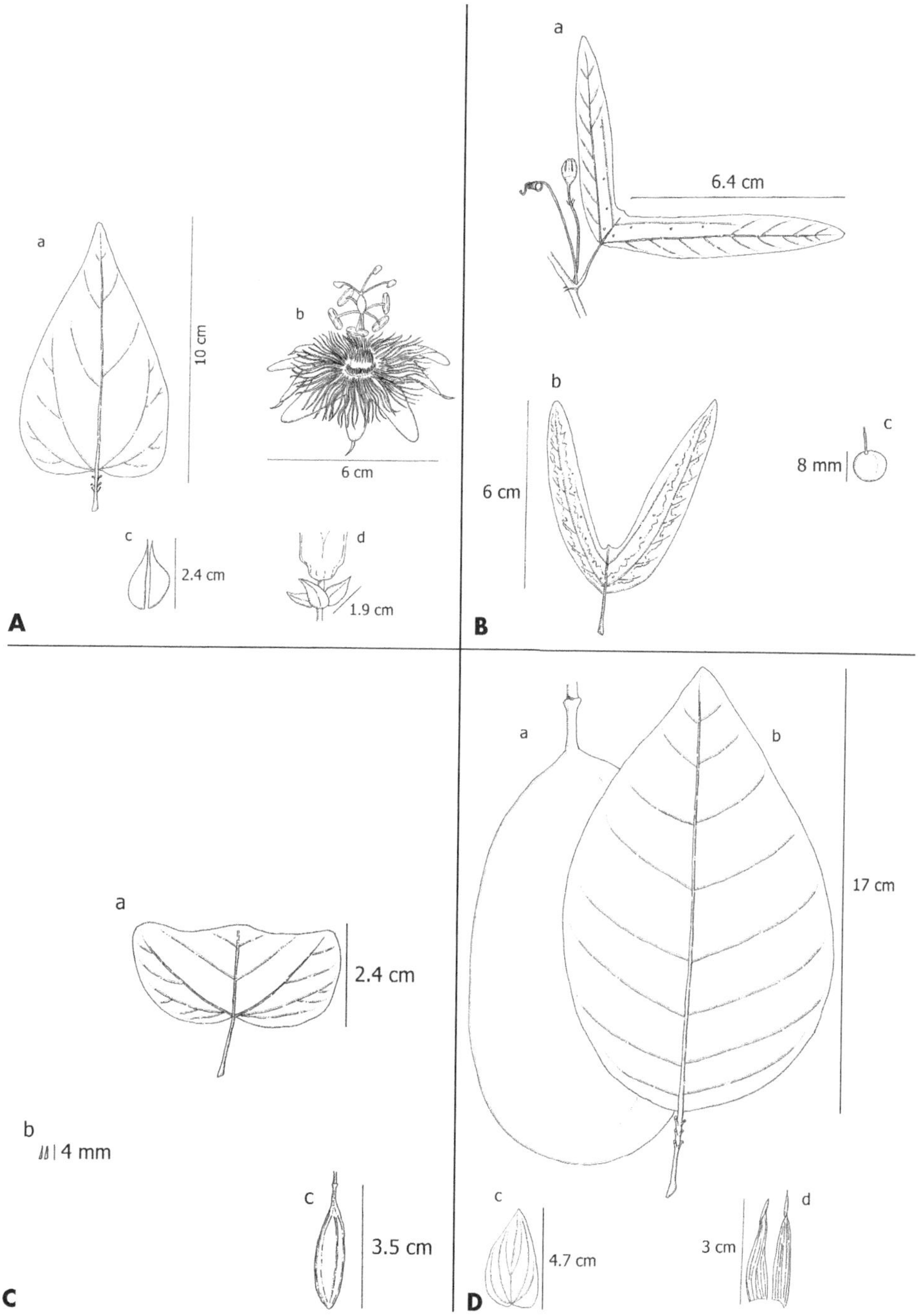

Fig. 7.9. A. *Passiflora oerstedii:* **(a)** leaf, **(b)** flower, **(c)** stipules, **(d)** floral bracts (Leaf and stipules based on Alvarado 4, CR; Floral bracts and flower based on Garwood et al. 748, CR). **B.** *Passiflora ornithoura:* **(a)** leaf, stipule, tendril, floral bracts, and flower bud; **(b)** leaf showing variegation; **(c)** immature fruit (Based on Smith 2136, US). **C.** *Passiflora pusilla:* **(a)** leaf, **(b)** stipules, **(c)** fruit (Based on Jiménez M. 2136, CR). **D.** *Passiflora quadrangularis:* **(a)** fruit, **(b)** leaf, **(c)** floral bract, **(d)** stipules (Leaf, stipules, and fruit based on Grayum et al. 4687, CR; Floral bract based on McDowell 1103, DUKE).

Phenology. Flowers from January through March, June, and September. Fruits in September.
Distribution. Mexico to Venezuela and Peru. In Costa Rica found from 450 to 1,500+ m.
Notes. *Passiflora oerstedii* var. *choconiana* is readily distinguished from the typical form by its deeply trilobate leaves.

30. *Passiflora ornithoura* Mast., Bot. Gaz. 16: 8. 1891. Figure 7.9B.

Synonymy. *Passiflora dicthophylla* Mast., Bot. Gaz. 16: 8. 1891.
Description. Vine; stems glabrous. Leaves simple, to 10 cm long and wide (frequently less), variable in shape, frequently strongly bilobate, glabrous above and below, glandular-ocellate below, palmately 3-veined from the base; lobes ovate to lanceolate; tip mucronate, acute to rounded; base cuneate, truncate, or rounded; margins entire; stipules setaceous, to 3 mm long; petioles eglandular. Tendrils simple, axillary. Peduncles paired, 1–2 cm long; bracts setaceous, to 1.7 mm long. Flowers to 1.5 cm wide; sepals oblong, to 8 mm long and 4 mm wide, not corniculate; petals white, to 5 mm long and 2.5 mm wide, linear; corona filaments generally uniseriate; ovary globose. Fruit bluish-black to purple, to 1 cm wide, globose. Seeds ovoid, to 2.5 mm long and 2 mm wide, transversely sulcate.
Phenology. Flowers and fruits in April in Guatemala. Although cited by Killip (1938), no fertile material from Costa Rica has been seen.
Distribution. Guatemala to Costa Rica. In Costa Rica found from 1,300 to 1,900 m.

31. *Passiflora pusilla* J.M. MacDougal, Ann. Mo. Bot. Gard. 75: 392. 1988. Figure 7.9C.

Description. Vine; stems angulate, glabrate to pubescent. Leaves simple, to 2.5 cm long and 4.5 cm wide, obovate, pubescent above and below, palmately 3-veined from the base, shallowly trilobate; lobes mucronate; tip obtuse, rounded, or truncate; base cordate; margins entire; stipules linear-triangular, to 5 mm long and less than 1 mm wide; petioles eglandular. Tendrils simple, axillary. Peduncles solitary; bracts absent. Flowers to 1 cm wide; sepals yellowish-green, to 2+ mm long, lanceolate, not corniculate; petals yellowish-green, 1.2 mm long, more or less ovate; corona filaments biseriate; inner filaments capillary, 1.2 mm long; outer filaments yellowish-green to yellow at apex, sometimes purple banded basally, to 6 mm long; ovary ellipsoid to obovoid, less than 1 mm long, puberulent. Fruit fusiform, to 7 mm long, hexagonal, 6-carinate. Seeds obovate, to 3 mm long and 2 mm wide, beaked, sometimes transversely sulcate.
Phenology. Flowering times unknown. Fruits in June.
Distribution. Nicaragua and Costa Rica. In Costa Rica found from sea level to 800 m.

32. *Passiflora quadrangularis* L., Syst. nat. ed. 10. 1248. 1759. Figure 7.9D.

Synonymy. *Passiflora sulcata* Jacq., Select. stirp. Amer. hist. 232. 1763; *Granadilla quadrangularis* (L.) Medik., Malvenfam. 97. 1787; *Passiflora tetragona* M. Roem., Fam. nat. syn. monogr. 2: 165. 1846; *P. macrocarpa* Linden ex Mast., Gard. Chron. 1869: 1012. 1869.
Description. Vine; stems glabrous, tetragonal with the angles usually winged. Leaves simple, to 20 cm long and 15 cm wide, unlobed, ovate to oblong-ovate, glabrous above and below, penni-veined; tip acute to acuminate; base rounded to subcordate; margins entire to serrulate; stipules ovate to ovate-lanceolate, to 4 cm long; petioles generally with 3 gland pairs (sometimes subopposite). Tendrils simple, axillary. Peduncles frequently 3-angulate, to 3 cm long, solitary; bracts cordate-ovate, to 5.5 cm long and nearly 4 cm

wide. Flowers to 12 cm wide; sepals ovate, to 4 cm long, white to pinkish or violet internally, greenish to reddish-green externally, corniculate; petals oblong-ovate to ovate-lanceolate, to 4.5 cm long and 2 cm wide, white to pinkish; corona filaments bluish or pinkish, mottled at apex and near middle, reddish-purple and white banded near base, frequently 5-seriate; outer filaments to 6 cm long; ovary ellipsoid to ovoid, to 6 mm long and 4 mm wide, glabrous to scattered puberulent. Fruit greenish to orange, to 30 cm long and 15 cm wide, oblong-ovoid. Seeds obcordate or suborbicular, to 1 cm long, flattened, reticulate (at least near center).

Phenology. Flowers in January and from March through October. Fruits in January and from April through August.

Distribution. Tropical America, but introduced throughout the world in the tropics. In Costa Rica found from sea level to 900+ m.

33. *Passiflora serratifolia* L., Sp. pl. 955. 1753. Figure 7.10A.

Synonymy. *Granadilla serratifolia* (L.) Medik., Malvenfam. 97. 1787; *Passiflora denticulata* Sessé & Moc., Fl. mexic. 227. 1887.

Description. Vine; stems terete, hirtellous. Leaves simple, to 12 cm long and 8 cm wide, unlobed, ovate to oblong-ovate, glabrate above, generally hirtellous below; tip acuminate; base rounded to subcordate; margins serrulate; stipules linear-subulate, to 7 mm long; petioles generally with 3 gland pairs (sometimes subopposite). Tendrils simple, axillary. Peduncles frequently solitary; bracts oblong-lanceolate, to 3 cm long, often pubescent. Flowers to 6 cm wide; sepals purplish, to 3 cm long and 0.8 cm wide, lanceolate, corniculate; petals purplish, to 2 cm long and 0.6 cm wide, oblong-lanceolate; corona filaments frequently multi-seriate; inner filaments purple and white banded, to 2 mm long; outer filaments white at apex, purplish toward base, to 3.5 cm long; ovary ovoid or ellipsoid, glabrous. Fruit yellow, to 9 cm long and 5 cm wide, subglobose to ovoid. Seeds flattened, to 7 mm long and 5 mm wide, reticulate, frequently 3-dentate at apex.

Phenology. Flowering times unknown. Fruits in August.

Distribution. Mexico to Costa Rica. In Costa Rica found from 100 to 1,980 m.

34. *Passiflora sexflora* Juss., Ann. Mus. Natl. Hist. Nat. 6: 110, t.37, f.2. 1805. Figure 7.10B.

Synonymy. *Passiflora pannosa* Sm., Cycl. 20: *Passiflora* No. 28. 1819; *P. capsularis* var. *geminifolia* DC., Prodr. 3: 325. 1828, orthographic error for *geminiflora*; *Meioperis pannosa* (Sm.) Raf., Fl. tellur. 4: 103. 1838; *Cieca pannosa* (Sm.) M. Roem., Fam. nat. syn. monogr. 2: 148. 1846; *Decaloba sexflora* (Juss.) M. Roem., Fam. nat. syn. monogr. 2: 164. 1846; *Passiflora floribunda* Lem., Fl. Serres 4: 335b. 1848; *P. capsularis* var. *geminiflora* DC. ex Mast. in Mart., Fl. Bras. 13: 590. 1872; *P. miraflorensis* Killip, J. Wash. Acad. Sci. 14: 109. 1924.

Description. Vine; stems subangulate to angulate, frequently densely pilose. Leaves simple, to 8 cm long and 10 cm wide, variable in shape, bilobate or trilobate with the middle lobe conspicuously reduced, pubescent above and below, not glandular-ocellate, palmately 3-veined from the base; tip acute to rounded; base cordate to subcordate or truncate; margins entire; stipules linear-subulate, to 5 mm long; petioles eglandular, pilose. Tendrils simple, axillary. Peduncles paired; bracts linear-lanceolate, to 5 mm long. Flowers to 3 cm wide; sepals white, to 1.5 cm long and about 0.4 cm wide, glabrous internally, hirsute externally, lanceolate, not corniculate; petals linear, to 1 cm long; corona filaments generally

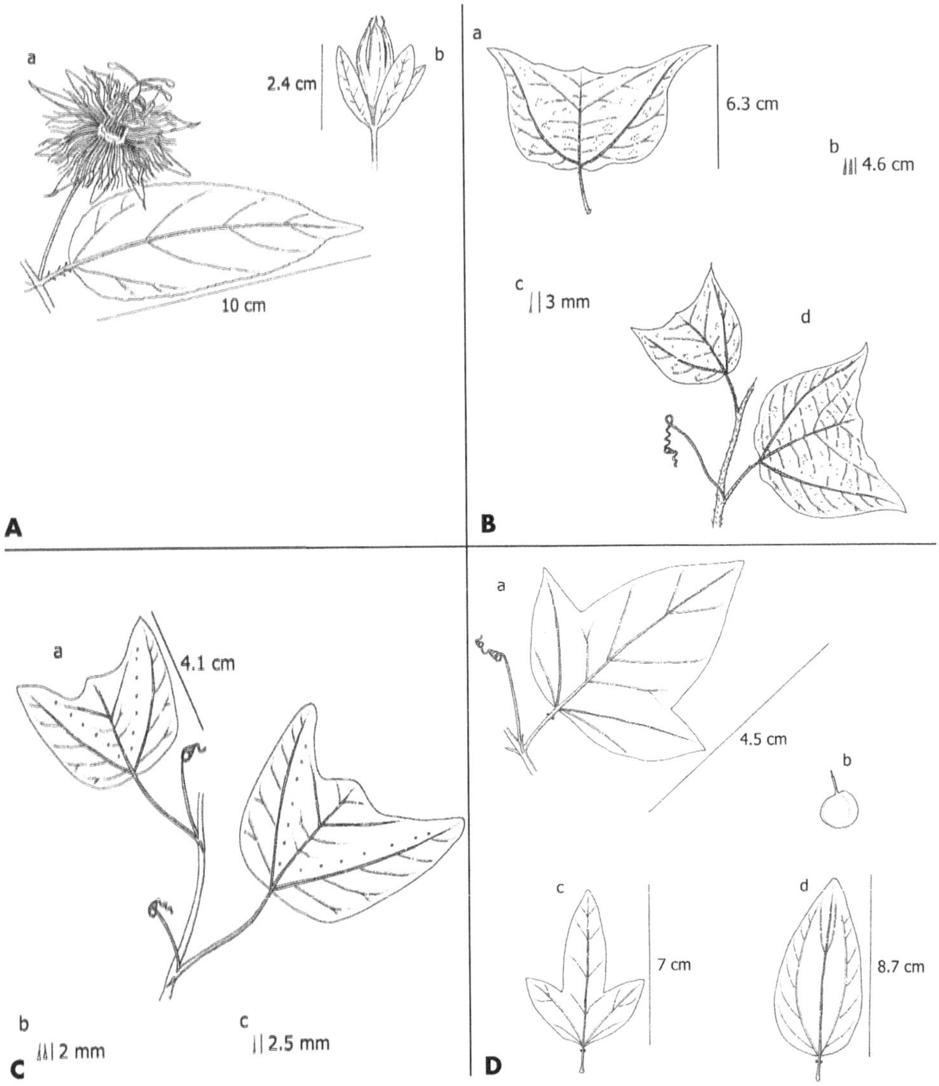

Fig. 7.10. A. *Passiflora serratifolia:* **(a)** leaf and inflorescence, **(b)** floral bracts and flower bud (Leaf and inflorescence based on Walker 1232, MO; Floral bracts and flower bud based on Cabrera 6933, MO). **B.** *Passiflora sexflora:* **(a)** leaf, **(b)** stipules, **(c)** floral bract, **(d)** leaves and tendril (Leaves, stipules, and tendril based on Krings 118, CR; Floral bract based on Smith 1170, DUKE). **C.** *Passiflora standleyi:* **(a)** leaves and tendrils, **(b)** stipules, **(c)** floral bract (Leaves, stipules, and tendrils based on Krings 122, CR; Floral bracts based on Taylor 3694, DUKE). **D.** *Passiflora suberosa:* **(a)** leaf, stipules, and tendril; **(b)** fruit; **(c & d)** leaf variation (Leaf, stipules, tendril, and fruit based on Tonduz 10903, US; Leaf variation based on Daniel 5650, US and Edwards 9, US).

biseriate; inner filaments purple, to 5 mm long; outer filaments white at apex, purplish toward base, to 1 cm long; ovary globose to ovoid, densely pubescent, tomentose, or villous. Fruit dark-green to purple, to 1 cm long and wide, globose, pubescent. Seeds obovate, at least 3 mm long, nearly 2 mm wide, flattened, transversely sulcate.

Phenology. Flowers from April through October. Fruits from April through July.

Distribution. Mexico to Colombia and Venezuela. In Costa Rica found from 1,200 to 2,025 m.

35. ***Passiflora standleyi*** Killip, J. Wash. Acad. Sci. 14: 110. 1924. Figure 7.10C.

Description. Vine; stems subangulate to angulate, glabrous. Leaves simple, to 5 cm long and 12 cm wide, variable in shape, bilobate or trilobate with the middle lobe conspicuously reduced, glabrous above and below, ocellate below, palmately 3-veined from the base; tip acute to obtuse; base rounded to subcuneate; margins entire; stipules linear, to 2 mm long; petioles eglandular. Tendrils simple, axillary. Peduncles to 3 cm long; bracts setaceous, to 3 mm long. Flowers to 4 cm wide; sepals lanceolate to ovate-lanceolate, to 1.5 cm long and about 0.4 cm wide, not corniculate; petals to 8 mm long; corona filaments generally biseriate; inner filaments white, to 5 mm long; outer filaments white and bluish-spotted at apex, bluish toward base, to 7 mm long; ovary subglobose to ovoid. Fruit purple to black, to 1.5 cm long and wide, globose. Seeds oblong, to 3 mm long and 2 mm wide, transversely sulcate.

Phenology. Flowering times unknown. Fruits in May and August.

Distribution. Guatemala to Costa Rica. In Costa Rica found from sea level to 1,800 m.

36. ***Passiflora suberosa*** L., Sp. pl. 958. 1753. Figure 7.10D.

Synonymy. The extensive synonymy for this species is beyond the scope of this guide; see Killip (1938).

Description. Vine; stems glabrate to pilose. Leaves simple, 4–8 cm long, 2.5–5 cm wide, 3-lobate, pubescent to glabrous above and below; tip acute; base rounded; margins entire; stipules elongate-triangular, to 5 mm long; petioles with a single gland pair near the apex. Tendrils simple, axillary. Peduncles solitary or paired in leaf axils, to 1.5 cm long, ebracteate. Flowers to 1.8 cm wide; sepals 5, about 1 cm long, cream, not corniculate; petals absent; corona filaments reddish at base; ovary ovoid. Fruit subglobose to globose, to 1.2 cm wide. Seeds reticulate, about 4 mm long, 2 mm wide.

Phenology. Flowering times unknown. Fruits in November.

Distribution. Hawaii, Florida, West Indies to Paraguay, Africa and Asia. In Costa Rica found from 1,200 to 1,530 m.

37. ***Passiflora talamancensis*** Killip, J. Wash. Acad. Sci. 12: 260. 1922. Figure 7.11A.

Description. Vine; stems angulate, frequently minutely puberulent. Leaves simple, to 14 cm long and 6.5 cm wide, elliptic to obovate, frequently shallowly trilobate at apex with the middle lobe dominant, glabrous above, glabrous or puberulent but always ocellate below, palmately 3-veined from the base; lobes mucronate; tip acute; base cuneate to narrowly rounded; margins entire; stipules linear-subulate, to 8 mm long; petioles eglandular, puberulent. Tendrils simple, axillary. Peduncles frequently paired; bracts setaceous, to 2 mm long. Flowers to 3.5 cm wide; sepals oblong, to 1.5 cm long and about 0.6 cm wide, white

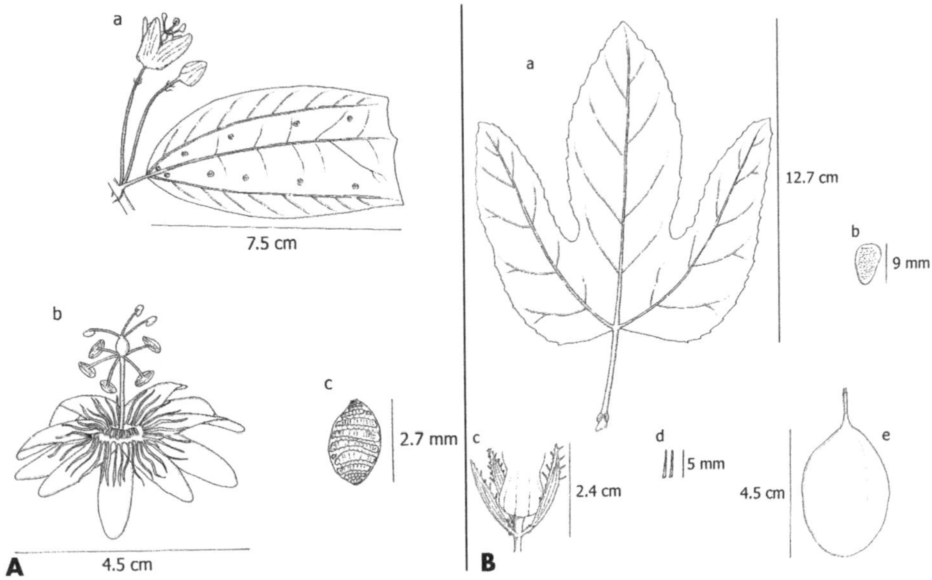

Fig. 7.11. A. *Passiflora talamancensis:* **(a)** leaf, stipules, and inflorescences; **(b)** flower, **(c)** seed (Leaf, stipules, and inflorescences based *on* Maas 7835, MO; Flower based on Mallet 22, CR; Seed based on Zamora 1482, CR). **B.** *Passiflora vitifolia:* **(a)** leaf, **(b)** seed, **(c)** floral bracts, **(d)** stipules, **(e)** fruit (Leaf, stipules, and seed based on Espinoza 142, CR; Floral bracts based on Robles 2733, CR).

internally, greenish externally, not corniculate; petals white, to 1 cm long and 0.2 cm wide, oblong; corona filaments generally biseriate; inner filaments to 2 mm long; outer filaments to 7 mm long; ovary globose or ellipsoid, densely pilosulose or woolly-tomentulose. Fruit purplish to black, to 1.2 cm long and wide, globose. Seeds ovate, to 5 mm long and about 2 mm wide.

Phenology. Flowers from October through March and in July. Fruits from November through February.

Distribution. Nicaragua to Panama. In Costa Rica found from sea level to 1,800 m.

Notes. *Passiflora talamancensis* is similar to *P. helleri* in general appearance. Leaves of *P. talamancensis* are generally fully twice as long as wide, but in *P. helleri* only 1.5 times as long as wide. The staminal filaments are also often shorter than 0.7 cm in *P. talamancensis*, whereas they are 1 cm or longer in *P. helleri.*

38. *Passiflora vitifolia* Kunth, Nov. Gen. sp. pl. 2: 138. 1817. Figure 7.11B.

Synonymy. *Passiflora sanguinea* Sm., Cycl. 26: *Passiflora* No. 45. 1819; *P. punicea* Ruíz & Pav. ex DC., Prodr. 3: 329. 1828; *Tacsonia sanguinea* (Sm.) DC., Prodr. 3: 334. 1828; *Macrophora sanguinea* (Sm.) Raf., Fl. tellur. 4: 103. 1838; *Passiflora vitifolia* var. *cassiquiarensis* M. Roem., Fam. nat. syn. monogr. 2: 181. 1846; *P. serrulata* var. *pubescens* Griseb., Bonplandia 6: 7. 1858; *P. servitensis* H. Karst., Linnaea 30: 163. 1859; *P. buchanani* Lem., Ill. hort. 14: pl. 519. 1867.

Description. Liana or vine; stems terete, densely reddish or golden tomentose. Leaves simple, to 14 cm long and 17 cm wide, 3-lobate, pubescent above and below; tip acute to acuminate; base cordate, subcordate, or truncate; margins serrate; stipules setaceous, to 5 mm long; petioles with 1 gland pair near or at base. Tendrils simple, axillary. Peduncles to 9 cm; bracts oblong to lanceolate, to 2.5 cm long, glandular-serrate or entire. Flowers reddish to scarlet, 14 cm wide; sepals lanceolate, to 8 cm long and 2 cm wide, somewhat puberulent externally, corniculate; petals to nearly 6 cm long and 1.5 cm wide; corona filaments 3-seriate; inner filaments paler in color, to 1 cm long; outer filaments red or yellow, to 2 cm long; ovary ellipsoid, minutely tomentose. Fruit ovoid, to 6 cm long and 3+ cm wide, puberulent. Seeds obcordate, to 5 mm long and 4 mm wide, flattened, reticulate.

Phenology. Flowers from July through April. Fruits from January through March and from July through September.

Distribution. Nicaragua to Venezuela and Peru. In Costa Rica found from 200 to 1,100+ m.

8
POLEMONIACEAE

The Polemoniaceae encompasses 18 genera and 300 species with a mostly temperate distribution (Moore 1978). In our area it is represented by five climbing species of *Cobaea*. The tubular flowers of *Cobaea* with long peduncules are usually unmistakable. Vegetatively, the genus is easily recognized by alternate, pinnately compound leaves with the terminal leaflets modified into tendrils. The only other family exhibiting similar leaf and tendril structures is Fabaceae, distinguished by either bipinnately compound leaves (*Entada*) or small, 4-many-jugate leaflets (*Vicia*). Unlike *Cobaea*, the Fabaceae often have swollen pulvini. Important references to our members of this family include Standley (1914), Gibson (1970), Austin (1971), and Prather (1999).

1. GENUS *COBAEA*

In our area five species of *Cobaea* are found. They are easily distinguished by the shape of the corolla.

KEY TO *COBAEA*

1. Corolla lobes ovate to orbicular
   2. Sepals triangular or lanceolate, less than 0.7 cm wide; leaflets shorter than 5 cm... *C. minor*, p. 121.
   2. Sepals ovate to suborbicular, frequently greater than 2 cm wide; leaflets 4–10 cm long... *C. scandens*, p. 122.
1. Corolla lobes linear-elongate
   3. Corolla lobes linear, uniformly tapering from the base... *C. gracilis*, p. 121.
   3. Corolla lobes ovate at base, abruptly contracted into an attenuate or acuminate tip
      4. Stems villous; calyx villous; corolla lobes (2.7) 3.8–5.9 cm long... *C. aschersoniana*, p. 119.
      4. Stems glabrous to puberulent, not villous; calyx glabrous to puberulent, not villous; corolla lobes 1.3–3.0 cm long... *C. lutea*, p. 121.

1. ***Cobaea aschersoniana*** Brand, Helios 21: 87. f.2. 1904. Figure 8.1A.

Synonymy. *Rosenbergia aschersonia* House, Muhlenbergia 4: 25. 1908.

Description. Vine, stems pubescent. Leaves pinnately compound; terminal leaflet modified into a branched tendril; lateral leaflets oblong to oblanceolate, 4–8 cm long, 2–3.5 cm wide, glabrate above, pubescent on the veins below; tip acute to acuminate; base asymmetrical and rounded or slightly cordate; margins entire. Inflorescence solitary or paired, axillary. Flowers with pedicels 14–20 cm long; sepals green, to 7 cm long; sepal lobes ovate at base, abruptly contracted into long-attenuate tips; corolla tube purplish-green (chartreuse?), to 8 cm long, campanulate, slightly longer than the sepals; stamens 5, densely woolly at base; style with branched stigmas. Fruit ellipsoid, to 5 cm long and 2 cm wide, glabrous, glaucous. Seeds 3 per locule, to 2.5 cm long and 1.1 cm wide.

Phenology. Flowers in January, February, and September. Fruits in April.

Distribution. Endemic to Costa Rica. Found from 1,600 to 2,800 m.

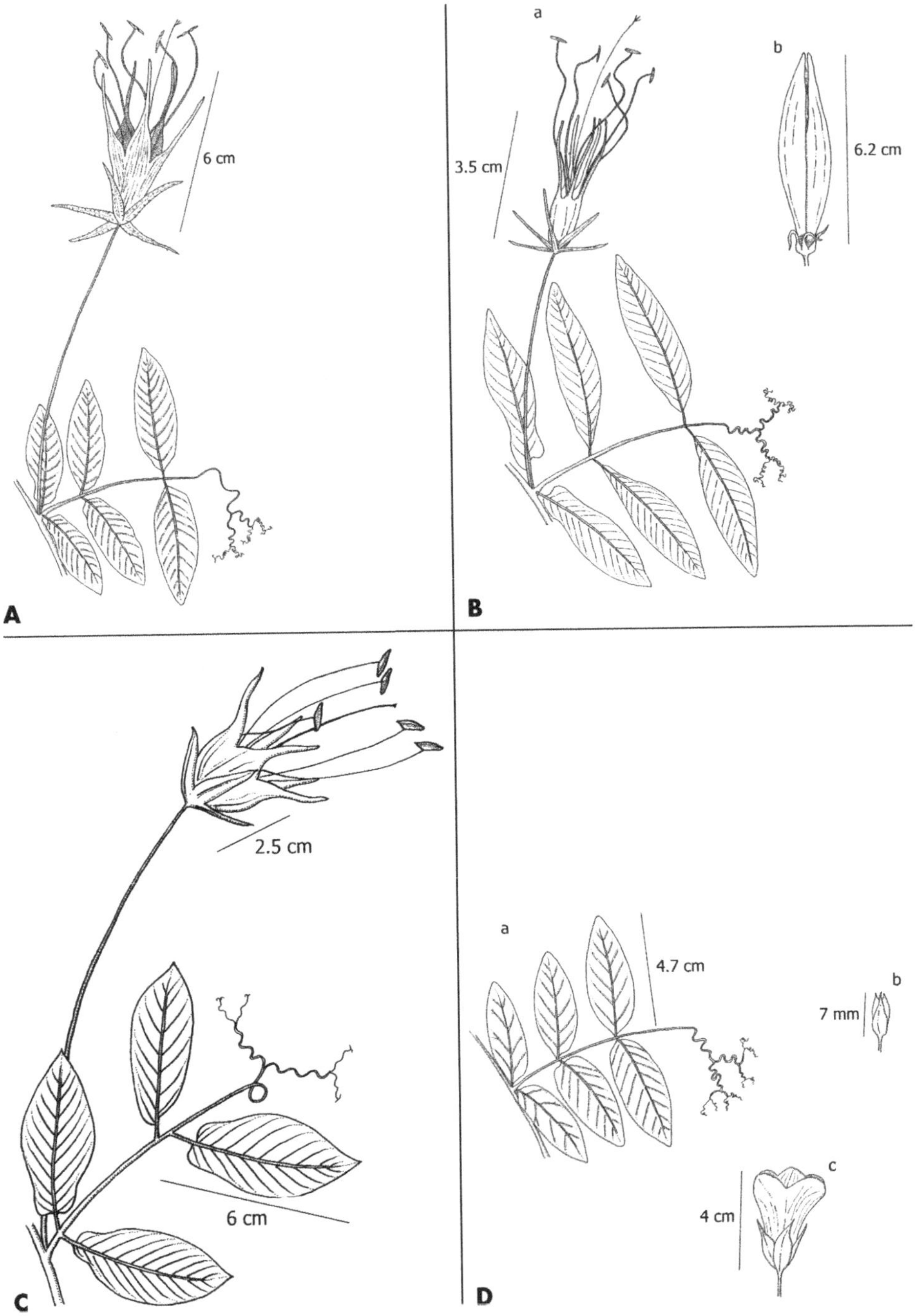

Fig. 8.1. A. *Cobaea aschersoniana:* leaf and inflorescence (Based on Gómez 19894, MO). **B.** *Cobaea gracilis:* **(a)** leaf and inflorescence, **(b)** fruit (Based on Grayum and Jacobs 3790, MO). **C**. *Cobaea lutea:* leaves and flower. **D.** *Cobaea minor:* **(a)** leaf, **(b)** flower bud (Based on Gómez et al. 21941, MO), **(c)** flower.

2. *Cobaea gracilis* (Oerst.) Hemsl., The Garden 17: 352. 1880. Figure 8.1B.

Synonymy. *Rosenbergia gracilis* Oerst., Vidensk. Meddel. Dansk Naturhist. Foren. Kjobenhavn 1856: 31. 1856; *Cobaea panamensis* Standl., Contr. U.S. Natl. Herb. 17: 452. 1914.

Description. Vine; stems somewhat 4-angulate, glabrate. Leaves pinnately compound; terminal leaflet modified into a tendril; lateral leaflets oblong to obovate, 3.5–7 cm long, 1.1–2.2 cm wide; tip acute to acuminate; base asymmetrically rounded or cordate; margins entire. Inflorescence solitary, axillary. Flowers with pedicels 9–20+ cm long; sepals green, to 2.5 cm long, linear-lanceolate; corolla tube yellow or greenish-yellow, to 1.7 cm long, more or less campanulate; corolla lobes to 3.5 cm long; stamens 5; style with branched stigmas. Fruit fusiform, to 6.5 cm long and 2 cm wide, glabrous.

Phenology. Flowers and fruits from February to April and in August.

Distribution. Costa Rica and Panama. In Costa Rica found from 700 to 1,670 m.

3. *Cobaea lutea* D. Don, Edinburg Phil. J. 10: 112. 1824. Figure 8.1C.

Synonymy. *Cobaea villosa* Standl., Contr. U.S. Natl. Herb. 17: 454. 1914; *Cobaea lutea* f. *villosa* (Standl.) Gibson, Fieldiana: Bot. 31: 354. 1968; *Cobaea viorna* Standl., Contr. U.S. Natl. Herb. 17: 453; *Cobaea lutea* f. *viorna* (Standl.) Gibson, Fieldiana: Bot. 31: 354. 1968; *Cobaea steyermarkii* Standl., Field. Mus. Bot. 22: 271. 1940.

Description. Liana; stems glabrous or puberulent at nodes. Leaves pinnately compound; terminal leaflet modified into a branched tendril; lateral leaflets 2.4–11 cm long, 1.1–6 cm wide, usually elliptic to elliptic-oblanceolate, glabrous to puberulent along the veins; tip acute to acuminate; base cuneate, truncate, or cordate; margins entire, minutely ciliate to long-ciliate at base. Inflorescence 1–5-flowered, axillary. Flowers with pedicels (8.2) 14.3–25 cm long; sepals green, to 2.8 cm long; corolla tube green, rarely greenish-yellow or purplish, campanulate, 1.7–3.7 cm long; corolla lobes broadly ovate, narrowing to long acuminate tips; stamens 5; style with branched stigmas. Fruit 2–5.8 cm long. Seeds winged, 1.6–2.1 cm long, 1.1–1.2 cm wide.

Phenology. Flowers and fruits from June to February (Prather 1999).

Distribution. Mexico to El Salvador, Costa Rica, and Peru. In Costa Rica, known from Alajuela, although this record may be a mistake (Prather 1999).

4. *Cobaea minor* M. Martens & Galeotti, Bull. Acad. Roy. Sci. Belgique 12: 276. 1845. Figure 8.1D.

Synonymy. *Rosenbergia minor* House, Muhlenbergia 4: 24. 1908.

Description. Vine; stems glabrous. Leaves pinnately compound, glabrous or with widely scattered hairs; terminal leaflet modified into a many-branched tendril; lateral leaflets oblong to narrowly elliptic, to 5 cm long (frequently 2–4 cm), 1.2–2 cm wide; tip acute; base rounded, truncate, or subcordate; margins entire. Inflorescence solitary, axillary. Pedicels glabrous; sepals green, to 2 cm long and 0.7 cm wide, lanceolate to narrowly ovate-lanceolate; corolla tube bluish to whitish-blue with purple streaks, to 3.5 (4) cm long, campanulate; corolla lobes ovate, broadly rounded at apex; stamens 5. Fruits not seen.

Phenology. Flowers in January, April, May, and August through October. Fruiting times unknown.

Distribution. Southern Mexico to Panama. In Costa Rica found from 2,000 to 3,200 m.

5. *Cobaea scandens* Cav., Icon. Pl. 1: 11. pl. 16, 17. 1791. Figure 8.2.

Synonymy. *Rosenbergia scandens* House, Muhlenbergia 4: 23. 1908.

Description. Vine; stems glabrous. Leaves pinnately compound, glabrous; terminal leaflet modified into a many-branched tendril; lateral leaflets oblong to oblanceolate, 4–10 cm long, 1.5–3 cm wide; tip acute to acuminate; base asymmetrical and rounded or slightly cordate; margins entire. Inflorescence solitary, axillary. Pedicels glabrous, to 18 cm long; sepals broadly ovate to suborbicular, to 2.5 cm wide; corolla tube purple or green or with purple stripes, to 4 cm long, campanulate, puberulent to glabrous externally; stamens 5, exserted about 1 cm beyond the tube; style with branched stigmas. Fruit ellipsoid, to 6 cm long and 2.5 cm wide, green, glabrous.

Phenology. Flowers in August. Fruits in September.

Distribution. Mexico to Peru. In Costa Rica found from 1,000 to 1,800 m.

Fig. 8.2. *Cobaea scandens:* **(a)** leaf and inflorescence, **(b)** fruit (Leaf and inflorescence based on Croat 586, MO; Fruit based on Hammel et al. 15546, MO).

9
POLYGONACEAE

Although cosmopolitan in distribution, the Polygonaceae is a largely north temperate family comprising about 30 genera and 750 species (Wright 1979). The family contains several genera with lianas (notably *Muehlenbeckia* and *Coccoloba*), but only *Antigonon* is tendrillate in our area. *Antigonon* is identified by tendrils that terminate inflorescences.

1. GENUS *ANTIGONON*

Both of our species of *Antigonon* are cultivated and have only occasionally escaped. They are primarily distinguished by the shape of the persistent perianth parts that often enclose the fruit. Burger (1983) indicated that in Central America both species are commonly called either *Bellisima* or *Confite*. Individually, *Antigonon guatemalense* is called *San Andres*, whereas *A. leptopus* is called *Flor de San Miguel.*

KEY TO *ANTIGONON*

1. Petioles usually shorter than 1 cm; floral bracts ovate, to 0.5 cm long; outer perianth segments to 3.5 cm long and wide, rounded at apex, cordate at base... *A. guatemalense*, p. 123.
1. Petioles usually longer than 2 cm; floral bracts lanceolate, to 0.3 cm long; outer perianth segments less than 1.5 cm long and 1 cm wide, acute at apex, subcordate at base... *A. leptopus*, p. 123.

1. ***Antigonon guatemalense*** Meisn., in DC. Prodr. 14: 184. 1856. Figure 9.1A.

Description. Vine or liana; stems densely brownish puberulent. Leaves simple, to 15 cm long and 10 cm wide, ovate to deltoid-ovate, puberulent to somewhat velutinous below; tip acute; base cordate; margins entire. Inflorescence frequently solitary, axillary (sometimes in a terminal panicle), to 23 cm long, terminating in a tendril; floral bracts broadly ovate, to 5 mm long, puberulent. Flowers rose-colored. Outer perianth segments to 3.5 cm long and wide; tip emarginate to rounded; base cordate. Fruit ovoid, to 10 mm long and 5 mm wide, brown to greenish, 3-ridged longitudinally near apex.

Phenology. Flowering times unknown. Fruits from August to December in Nicaragua.

Distribution. Probably native to Guatemala, but widely cultivated in the tropics and sometimes escaping. In Costa Rica found in gardens or thickets from sea level to 1,500 m.

2. ***Antigonon leptopus*** Hook. & Arn., Bot. Beechey Voy. 308, pl. 69. 1839–1840. Figure 9.1B.

Description. Vine or liana; stems variously puberulent. Leaves simple, to 15 cm long and 10 cm wide, ovate to somewhat sagittate, variously puberulent below; tip acute to acuminate; base cordate to subcordate; margins entire. Inflorescence frequently solitary, to 25 cm long, axillary (sometimes in a terminal panicle), terminating in a tendril; floral bracts lanceolate, to 3 mm long, sparsely puberulent. Flowers pinkish, violet, or rose. Outer per-

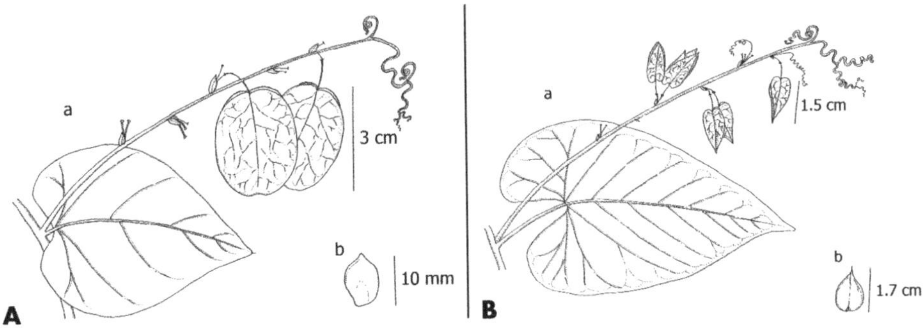

Fig. 9.1. A. *Antigonon guatemalense:* **(a)** leaf and infructescence, **(b)** fruit (Leaf and infructescence based on Moreno 4996, MO; Fruit based on Neill 2382, MO). **B.** *Antigonon leptopus:* **(a)** leaf and infructescence (showing bracts and terminating tendril), **(b)** fruit (Based on Molina R. 33980, MO).

ianth segments to 1.5 cm long and 1 cm wide; tip acute; base more or less subcordate. Fruit 3-angled and ovoid, to 10 mm long and 5 mm wide, brownish, 3-ridged longitudinally near apex.

Phenology. Flowers in January, February, May. Fruits in February.

Distribution. Probably native to Mexico, but widely cultivated in the tropics and sometimes escaping. In Costa Rica found in gardens or thickets from sea level to 1,200 m.

10
RHAMNACEAE

With 58 genera and about 900 species of cosmopolitan distribution (Jones 1978), the Rhamnaceae is represented in Costa Rica by only one tendrillate genus, *Gouania*. The only other genus of climbers in this family in Costa Rica, *Sagaretia*, has thorns. See Howard (1979) for additional information.

1. GENUS *GOUANIA*

Four species of *Gouania* occur in our area. One additional species, *G. colombiana* Suess, is known, but has not been collected in the mountains. The genus is identified by simple tendrils borne at the tip of short branches in the axils of leaves or at the base of inflorescences. The tendrils are quite conspicuous, and due to their uniplanar, curling nature have been described as resembling a butterfly proboscis or a watch spring. Leaf margins range from entire to serrate to glandular-serrate. Glands may also be present on the bases of leaves. In addition, the hooded stamens are unique among our tendrillate families.

KEY TO *GOUANIA*

1. Leaves densely white-tomentose or tomentulose below; hairs densely matted, often obscuring the leaf surface; leaf margins usually entire but sometimes shallowly crenate to serrulate... *G. hypoglauca*, p. 127.
1. Leaves glabrous or variously pubescent below; hairs not densely matted, not obscuring the leaf surface; leaf margins entire, crenate, or serrate
    2. Leaves suborbicular to orbicular, sometimes broadly elliptic, velutinous below; tip obtuse; margins crenate. Seed-bearing portion of fruit hirsute or hirtellous... *G. eurycarpa*, p. 125.
    2. Leaves elliptic, sometimes broadly so, variously pubescent to glabrous below; tips acute or short-acuminate; margins serrate, dentate, crenate, or subentire. Seed-bearing portion of fruit glabrous or with very scattered, more or less appressed hairs
        3. Fruit wing attached to the sides as well as above and/or below the seed-bearing portion. Floral disc glabrous throughout, or minutely appressed-puberulent... *G. lupuloides*, p. 127.
        3. Fruit wing attached only to the sides of the seed-bearing portion. Floral disc somewhat pubescent with long and conspicuous trichomes near the emergence of the styles... *G. polygama*, p. 127.

1. *Gouania eurycarpa* Standl., Publ. Field Mus. Nat. Hist., Bot. Ser. 4: 315. 1929. Figure 10.1A.

Description. Liana or climbing shrub; stems pubescent. Leaves simple, to 7 cm long and 5.5 cm wide, suborbicular to orbicular, sometimes broadly elliptic, sparsely pubescent above, velutinous below; tip obtuse; base obtuse, subcordate, or cordate; margins crenate; stipules broadly ovate, suborbicular, or reniform; petioles densely pubescent, 4–6 mm long. Tendrils simple, arising on short shoots in axils of leaves. Inflorescence thyrsiform.

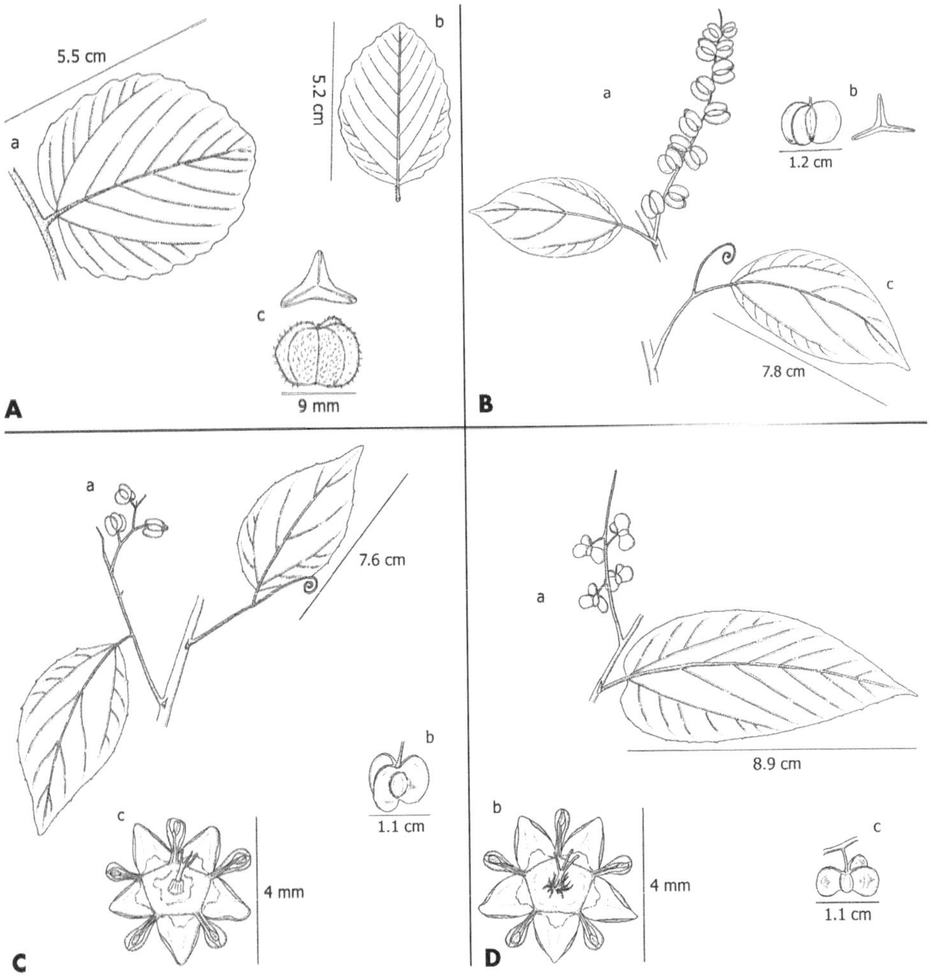

Fig. 10.1. A. *Gouania eurycarpa:* **(a)** leaf, **(b)** leaf variation, **(c)** fruit (Based on Berrocal and Sánchez 107, MO). **B.** *Gouania hypoglauca:* **(a)** leaf and infructescence, **(b)** fruit, **(c)** leaf and tendril (Based on Correa and Dressler 986, DUKE). **C.** *Gouania lupuloides:* **(a)** leaf, tendril, and infructescence; **(b)** fruit; **(c)** flower (Leaf, tendril, infructescence, and fruit based on Cabrera 15276, MO; Flower based on Gentle 236, F). **D.** *Gouania polygama:* **(a)** leaf and infructescence, **(b)** flower, **(c)** fruit (Leaf, infructescence, and fruit based on Lent 284, MO; Flower based on Chacón 718, F).

Flowers in sessile or subsessile glomerules, subtended by a single, elongate triangular bract; calyx 5-lobed; calyx lobes triangular, to 1 mm long, pubescent; corolla 5-lobed; stamens 5; style trifurcate. Fruit a 3-winged schizocarp, to 7 mm long and 10 mm wide; winged along the sides of the seed-bearing portion as well as above and below; seed-bearing portion hirsute or hirtellous. Seeds 3.

Phenology. Flowering times unknown. Fruits in July and from December through February.

Distribution. Southern Mexico and Belize to Costa Rica. In Costa Rica known only from Guanacaste Province. Probably mostly found at low elevations, but included here based on one collection from 500 to 1,000 m (Berrocal and Sanchez 107, MO).

Notes. The Guanacaste population described above may differ from other populations in Central America. In particular, leaf shapes and margins vary considerably. All Central American specimens examined, however, had hirsute, seed-bearing portions of the fruit.

2. *Gouania hypoglauca* Standl. Publ. Field. Mus. Nat. Hist., Bot. Ser. 22: 89. 1940. Figure 10.1B.

Description. Liana or climbing shrub; young stems pubescent. Leaves simple, to 9 cm long and 4.5 cm wide, elliptic to elliptic-ovate, sparingly pubescent above, densely white-tomentose or tomentulose below, hairs frequently matted and obscuring the surface; tip acute; base obtuse to rounded, rarely subcordate; margins entire to subcrenate; stipules small, caduceus; petioles pubescent, to 1 cm long. Tendrils simple, arising on short shoots in axils of leaves. Inflorescence thyrsiform. Flowers in sessile or subsessile glomerules, subtended by a single, narrowly deltoid bract; calyx 5-lobed; calyx lobes deltoid, to 1 mm long, pubescent; corolla tube pubescent, to 2.5 mm long, 5-lobed; corolla lobes deltoid, to 1 mm long; stamens 5; nectariferous disk bilobed, located between stamens, pubescent; style trifurcate. Fruit a 3-winged schizocarp, to 1.3 cm wide; wings attaching to the sides and apex and/or base of the seed-bearing portion; seed-bearing portion glabrous. Seeds 3.

Phenology. Flowers from April to June. Fruits in July and September.

Distribution. Costa Rica and Panama. In Costa Rica found from sea level to 750 m.

3. *Gouania lupuloides* (L.) Urb., Symb. antill. 4: 378. 1910. Figure 10.1C.

Synonymy. *Banisteria lupuloides* L., Sp. pl. 427. 1753; *Rhamnus domingensis* Jacq., Enum. syst. pl. 17. 1760; *Gouania domingensis* L., Sp. pl. ed. 2. 1763.

Description. Liana or climbing shrub; young stems pubescent. Leaves simple, to 9 cm long and 4.5 cm wide, elliptic to elliptic-ovate, sparsely pubescent above, glabrous to pubescent below; tip acute; base shallowly cordate to rounded; margins serrate or glandular-serrate; stipules small, caducous; petioles pubescent, to 2 cm long. Tendrils simple, arising on short shoots in the axils of leaves. Inflorescence thyrsiform. Flowers in sessile or subsessile glomerules, subtended by a single, deltoid bract; calyx 5-lobed; calyx lobes deltoid, to 1 mm long, pubescent; corolla 5-lobed; stamens 5; nectariferous disk unilobed or bilobed, located between stamens, glabrous throughout; styles trifurcate. Fruit a 3-winged schizocarp, to 1.2 cm wide; wings attached to the sides and apex and/or base of the seed-bearing portion; seed-bearing portion glabrous or with scattered, more or less appressed hairs. Seeds 3.

Phenology. Flowers in September and October. Fruiting times unknown.

Distribution. Mexico to northern South America; West Indies. In Costa Rica found from sea level to 1,600 m.

Notes. This species is known variously as *xomac*, *chevez-ac*, or *xpajuyic* in Mayan; *jaboncillo*, *canillo*, or *pie de pava* in Guatemala; and *limpia-dientes* in Central America (Standley and Steyermark 1949).

4. *Gouania polygama* (Jacq.) Urb., Symb. Antill. 4: 378. 1910. Figure 10.1D.

Synonymy. *Rhamnus polygamus* Jacq., Enum. syst. pl. 17. 1760; *G. tomentosa* Jacq., Sel. Stirp. Amer. 263. 1763.

Description. Liana or climbing shrub; young stems pubescent. Leaves simple, to 9 cm long and 4.5 cm wide, elliptic to elliptic-ovate, sparsely pubescent above, glabrous to pubescent below; tip acute; base shallowly cordate to rounded; margins serrate, glandular-serrate or crenate; stipules small, caducous; petioles pubescent, to 2 cm long. Tendrils simple, arising on short shoots in the axils of leaves. Inflorescence thyrsiform. Flowers in sessile or subsessile glomerules, subtended by a single, deltoid bract; calyx 5-lobed; calyx lobes deltoid, to 1 mm long, pubescent; corolla 5-lobed; corolla lobes deltoid, to 1 mm long; stamens 5; nectariferous disk unilobed or bilobed, located between stamens, somewhat pubescent near the emergence of the styles; styles trifurcate. Fruit a 3-winged schizocarp, to 1.3 cm wide; winged only along the sides of the seed-bearing portion; seed-bearing portion glabrous or with scattered appressed hairs. Seeds 3.

Phenology. Flowers in January, March, June, July, November, and December. Fruits from December through March.

Distribution. Florida, West Indies, and Mesoamerica to northern South America. In Costa Rica found from sea level to 1,500 m.

11
SAPINDACEAE

A family of trees, shrubs, lianas, and herbaceous vines, the Sapindaceae comprises about 150 genera with 2,000 species of mostly tropical and subtropical distribution (Morley 1978). The family is represented in Costa Rica by 6 climbing genera: *Allosanthus*, *Cardiospermum*, *Paullinia*, *Serjania*, *Thinouia*, and *Urvillea*. However, only *Cardiospermum*, *Paullinia*, and *Serjania* are found in our area.

Vegetatively, climbers in the Sapindaceae are often exceedingly similar. The fruit is the most reliable character to distinguish the genera. Floral characters are of little value, as they frequently overlap. Nonetheless, the following characters are provided, because they reduce the possibilities. *Cardiospermum* is generally distinct from the other genera by its herbaceous habit. In our area, only *Paullinia* contains species with trifoliolate leaves. Stipules over 5 mm long usually indicate *Paullinia* (especially *P. bracteosa* and *P. ingifolia*). Both *Serjania valerii* and the lowland species *Paullinia rugosa* have broad, lacerate or fringed stipules, but can easily be separated by wood characters. Whereas the majority of *Paullinia* species have pinnately compound leaves, only 2 species of *Serjania* share this character (*S. valerii* and *S. lobulata*). Thorns or prickles are present only in *Serjania mexicana*.

Cross-sections of mature stems can also be helpful in this family. In our area, stems characterized by eight or more vascular cylinders arranged in a circle around the central vascular cylinder are found only in *Serjania caracasana*. For additional help with identification, table 11.1 groups most of our climbers in this family according to vegetative characters. *Serjania hispida* and *Paullinia itayensis* vel. sp. aff. are excluded as their wood characters are unknown. All species found above 700 m also occur at lower elevations. Additional information is found in Radlkofer (1931–1934), Standley and Steyermark (1949), and Croat (1976, 1978).

Our treatment differs somewhat from the draft treatment of Morales (2000) in that we maintain *Paullinia ingifolia* and *P. grandifolia* as separate species (distinguished by stipular differences), but place *P. trisulca* in synonymy under *P. pterocarpa*. Morales (2000) maintains the latter as separate species, distinguishing it primarily by the length of the inflorescence (less than 1.7 cm for *P. pterocarpa* and 2–8 cm for *P. trisulca*). Morales also cites *P. hymenobractea* from our area (based on Grayum and Hammel 8441, MO), although this specimen apparently lacks fruit, which is usually required for accurate generic placement. As placement of this species in *Paullinia* has been questioned by others, we have excluded it pending further study.

Table 11.1. Costa Rican Climbers in the Sapindaceae Grouped by Vegetative Characters

Character	Species
Leaves trifoliolate; wood cross-section with one vascular cylinder	*Paullinia austin-smithii* (petioles not winged) *Paullinia cururu* (petioles slightly to moderately winged) *Paullinia talamancensis* (petioles not winged) • *Paullinia granatensis* (capsules densely echinate) • *Paullinia turbacensis* (stems with dense, raised lenticels) • *Thinouia myriantha* (fruit samaroid; seed-bearing portion basal) • *Urvillea ulmacea* (fruit membranous, somewhat inflated)
Leaves trifoliolate; wood cross-section with 3–10 peripheral vascular cylinders surrounding a central one	*Paullinia alata* • *Serjania circumvallata* • *Serjania grandis*
Leaves variously compound; wood cross-section with 3–5 peripheral vascular cylinders surrounding a central one	*Paullinia alata* (leaves trifoliolate or pinnately 5-foliolate) *Serjania atrolineata* (leaves biternate) *Serjania lobulata* (leaves pinnately 5-foliolate) *Serjania paniculata* (leaves biternate) *Serjania paucidentata* (leaves biternate) *Serjania rhombea* (leaves biternate) *Serjania valerii* (leaves pinnately 5-foliolate) • *Paullinia pinnata* (leaves pinnately 5-foliolate) • *Serjania goniocarpa* (leaves biternate) • *Serjania grandis* (leaves trifoliolate) • *Serjania macrocarpa* (leaves biternate) • *Serjania schiedeana* (leaves pinnately 5-foliolate) • *Serjania triquetra* (leaves biternate)
Leaves trifoliolate or biternate; wood cross-section with 8–10 peripheral vascular cylinders surrounding a central one	*Serjania caracasana* (leaves biternate) • *Serjania circumvallata* (leaves trifoliolate) • *Serjania decapleuria* (leaves biternate) • *Serjania pyramidata* (leaves biternate)
Leaves variously compound; wood cross-section with one vascular cylinder	*Paullinia austin-smithii* (leaves trifoliolate) *Paullinia bracteosa* (leaves 2–3 jugate; lowest leaflet pair simple) *Paullinia brenesii* (leaves biternate) *Paullinia costaricensis* (leaves biternate) *Paullinia cururu* (leaves trifoliolate)

Table 11.1. *(continued)*

Character	Species
	Paullinia faginea (leaves pinnately 5-foliolate) *Paullinia fuscescens* (leaves biternate) *Paullinia grandifolia* (leaves 3-jugate; lowest leaflet pair trifoliolate) *Paullinia ingifolia* (leaves 3-jugate; lowest leaflet pair trifoliolate) *Paullinia macrocarpa* (leaves pinnately 5-foliolate) *Paullinia mallophylla* (leaves 3–5-jugate; lowest leaflet pair trifoliolate) *Paullinia pterocarpa* (leaves pinnately 5-foliolate) *Paullinia pterophylla* (leaves 3–4-jugate; lowest leaflet pair simple) *Paullinia talamancensis* (leaves trifoliolate) *Serjania acuta* (leaves biternate) *Serjania membranacea* (leaves biternate) *Serjania mexicana* (stems armed; leaves biternate) *Serjania racemosa* (leaves biternate) *Serjania rufisepala* (leaves biternate) • *Paullinia clavigera* (leaves pinnately 5-foliolate) • *Paullinia costata* (leaves pinnately 5-foliolate) • *Paullinia dasystachya* (leaves pinnately 5-foliolate) • *Paullinia fibrigera* (leaves pinnately 5-foliolate) • *Paullinia fraxinifolia* (leaves pinnately 5-foliolate) • *Paullinia granatensis* (leaves trifoliolate) • *Paullinia rugosa* (leaves pinnately 5-foliolate) • *Paullinia serjaniifolia* (leaves 3-jugate; lowest leaflet pair trifoliolate to pinnately 5-foliolate) • *Paullinia subnuda* (leaves pinnately 5-foliolate) • *Paullinia turbacensis* (leaves trifoliolate) • *Serjania cornigera* (leaves biternate) • *Thinouia myriantha* (leaves trifoliolate) • *Urvillea ulmacea* (leaves trifoliolate)

• Species are found in lowlands.

KEY TO SAPINDACEAE

1. Fruit a schizocarp, separating into 3-winged mericarps... *Serjania*, p. 142.
1. Fruit a capsule, thick-walled or bladderlike
   2. Capsule inflated, bladderlike, not winged; leaves biternate; plant herbaceous; seeds lacking a fleshy arillode... *Cardiospermum*, p. 132.
   2. Capsule relatively thick-walled, winged or not; leaves biternate or not; plant woody; seeds with a fleshy arillode... *Paullinia*, p. 132.

1. GENUS *CARDIOSPERMUM*

Our single species of *Cardiospermum* is a herbaceous vine with biternate leaves and inflated, bladderlike capsules. Our only other genus (*Urvillea*) with membranous fruits grows in the lowlands, and is readily identified by its often woody stem and trifoliolate leaves. *Paullinia* and *Serjania* also have species with biternate leaves, but can be distinguished by their woody growth and noninflated fruits.

1. ***Cardiospermum grandiflorum*** Sw., Prodr. 64. 1788. Figure 11.1A.

Description. Vine; stems angulate, frequently 5–6-ribbed, pubescent to glabrate. Leaves biternate; leaflets ovate to oblong-ovate, 5–9 cm long, 2–4 cm wide, scattered pubescent to glabrate above and below; tip acuminate to obtuse; base attenuate; margins serrate to dentate. Tendrils bifid, axillary. Inflorescence a thyrse, axillary. Flowers to 8 mm long; sepals glabrous; petals white. Fruit ellipsoid to obovoid, 5–6 cm long, 3-angled, inflated and bladderlike, membranous, not winged, glabrous. Seeds globose, to 7 mm wide; hilum distinctly narrower than the seed, round.

Phenology. Flowers and fruits from November through April.

Distribution. Found throughout tropical America and tropical Africa. In Costa Rica found from sea level to 1,000 m.

2. GENUS *PAULLINIA*

Paullinia typically contains canopy lianas. Vegetatively, they are often difficult to distinguish from *Serjania*. Both genera exhibit bifid tendrils and frequently white latex. Generic determination is most easily based on fruit. *Paullinia* generally has thick-walled, dehiscent capsules, whereas *Serjania* has schizocarps that separate into 3-winged, indehiscent mericarps. In addition, seeds of *Paullinia* are black, shiny, and contain an arillode. Seeds of *Serjania* are not arillate and have a woody testa. In flower, both genera are very similar. Each typically has five sepals, four petals, a floral disc formed into four glands, eight stamens, and trifid or triparted styles.

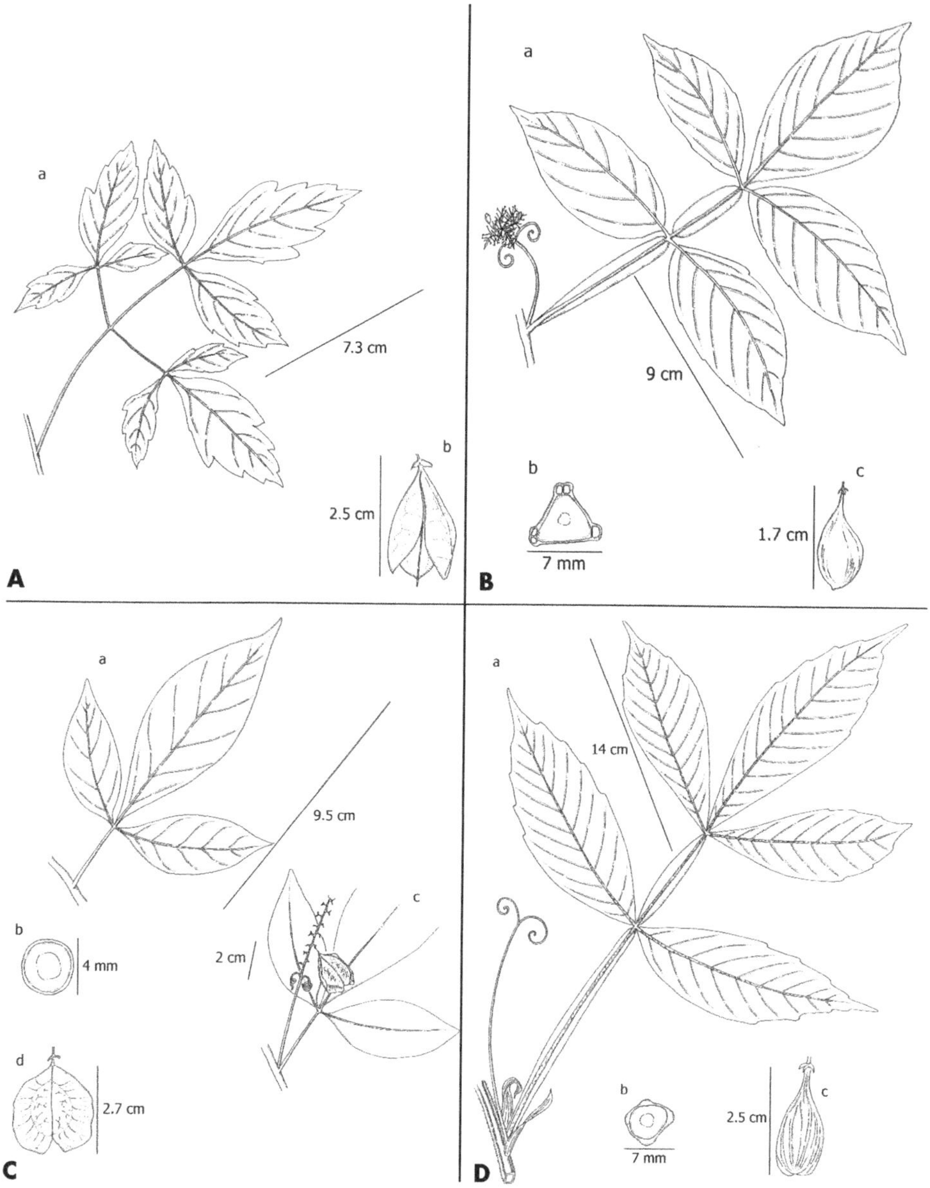

Fig. 11.1. A. *Cardiospermum grandiflorum:* **(a)** leaf, **(b)** fruit (Based on Croat 44425, MO). **B.** *Paullinia alata:* **(a)** leaf and inflorescence, **(b)** stem cross-section showing five vascular cylinders surrounding the central vascular cylinder, **(c)** fruit (Leaf and inflorescence based on Krings 140, CR; Stem cross-section and fruit based on Hunter and Allen 25, MO). **C.** *Paullinia austin-smithii:* **(a)** leaf, **(b)** stem cross-section showing a single vascular cylinder, **(c)** infructescence, **(d)** fruit (Based on Morales 1248, MO). **D.** *Paullinia bracteosa:* **(a)** leaf, stipules, and tendril; **(b)** stem cross-section; **(c)** fruit (Leaf, stipules, and tendril based on Krings 145, CR; Stem cross-section and fruit based on Liesner 2063, US).

KEY TO *PAULLINIA*

1. Leaves ternately compound (trifoliolate or biternate)
    2. Leaves biternate
        3. Fruit not winged... *P. costaricensis*, p. 136.
        3. Fruit winged
            4. Wings of fruit nearly 1.5 times longer than the seed-bearing portion; pedicels 3–5 mm long; floral bracts to 3.5 mm long... *P. brenesii*, p. 136.
            4. Wings of fruit as long as or shorter than the seed-bearing portion; pedicels ca. 2 mm long; floral bracts to 1.7 mm long... *P. fuscescens*, p. 138.
    2. Leaves trifoliolate
        5. Petioles winged
            6. Stems triangular; wood cross-section with 3 peripheral vascular cylinders surrounding the central one; inflorescence umbelloid, to 2 cm long, glomerulate... *P. alata*, p. 135.
            6. Stems terete; wood cross-section with a single vascular cylinder; inflorescence a solitary thyrse, 1.7–3.5 cm long... *P. cururu*, p. 136.
        5. Petioles not winged
            7. Fruits winged... *P. austin-smithii*, p. 135.
            7. Fruits not winged
                8. Fruits conspicuously curved... *P. cururu*, p. 136.
                8. Fruits suborbicular... *P. talamancensis*, p. 142.
1. Leaves pinnately compound
    9. Leaves 3–5-jugate; lowest pair of leaflets trifoliolate to pinnate
        10. Fruit winged; inflorescence glomerulate, to 2 cm long... *P. mallophylla*, p. 140.
        10. Fruit not winged; inflorescence not glomerulate, longer than 4 cm
            11. Stipules narrowly triangular, not clasping the stem, to 8 mm long and 2 mm wide... *P. grandifolia*, p. 138.
            11. Stipules ovate to broadly lanceolate, clasping the stem, 1.5–4 cm long, 6–8 mm wide... *P. ingifolia*, p. 138.
    9. Leaves 2–4-jugate; lowest pair of leaflets simple
        12. Stipules 2–5 cm long; wood cross-section with a single vascular cylinder... *P. bracteosa*, p. 135.
        12. Stipules shorter than 2 cm (if longer, leaf rachis not winged); wood cross-section with a single vascular cylinder or with peripheral vascular cylinders
            13. Fruit to 7 cm long (excluding stipe), smooth, not winged, stipitate; stipe to 2 cm long... *P. macrocarpa*, p. 140.
            13. Fruit less than 3 cm long (excluding stipe, if any), smooth or not, winged or not, stipitate or not
                14. Leaves 3- or more jugate; petiole and rachis winged... *P. pterophylla*, p. 142.
                14. Leaves 2-jugate; petiole and rachis winged or not
                    15. Wood cross-section with 3–6 peripheral vascular cylinders surrounding the central one; stipules less than 5 mm long; petiole and rachis winged; inflorescence densely glomerulate... *P. alata*, p. 135.
                    15. Wood cross-section with a single vascular cylinder; stipules 8–15 mm long; petiole and rachis not winged; inflorescence densely glomerulate or not
                        16. Fruit not winged; inflorescence solitary, 5.5–12 cm long; stipules to 10 mm long; stem, petiole, and rachis densely ferruginously hirtellous-tomentose or tomentulose... *P. faginea*, p. 136.

16. Fruit 3-winged; inflorescence glomerulate (rarely solitary), 3–5.5 cm long; stipules 8–15 mm long; stem, petiole and rachis glabrous... *P. pterocarpa*, p. 140.

1. *Paullinia alata* (Ruíz & Pav.) G. Don, Gen. Syst. 1: 660. 1831. Figure 11.1B.

Synonymy. *Semarillaria alata* Ruíz & Pav., Fl. peruv. 4: tab. 340. 1802; *Paullinia rhizantha* Poepp. & Endl., Nov. Gen. sp. pl. 3: 36, pt. 243. 1844; *P. fasciculata* Radlk., Monogr. Paullinia 124. 1895-1896; *P. metensis* Cuatrec., Revista Acad. Colomb. Ci. Exact. 4: 347. 1941.

Description. Liana; stems three-angled, villous when young, soon glabrate; wood cross-section typically with three peripheral vascular cylinders surrounding the central cylinder; peripheral vascular cylinders often splitting and appearing as 5–6 peripheral vascular cylinders. Leaves trifoliolate or pinnately 5-foliolate; leaflets elliptic to obovate, 3–19 cm long, 3–10 cm wide, glabrous above and below except pubescent on midrib above and in vein axils below; tip acuminate; margins crenate to serrate-dentate; stipules to 3 mm long; petioles and rachis winged. Tendrils bifid, axillary. Inflorescence glomerulate-fasciculate, axillary (or cauliflorous), thyrse, sparsely villous. Sepals glabrous, to 3 mm long; petals white to greenish-white. Fruit red, to 2 cm long, pyriform, not winged, glabrous. Seeds 1–3, ellipsoid, somewhat flattened.

Phenology. Flowers in January, May, and June. Fruits in March.

Distribution. Guatemala to Bolivia. In Costa Rica found from sea level to 1,500 m.

2. *Paullinia austin-smithii* Standl., Publ. Field Mus. Nat. Hist., Bot. Ser. 18: 1558. 1938. Figure 11.1C.

Description. Liana; stems triangular, 3-sulcate, or terete, glabrous; wood cross-section with a single vascular cylinder. Leaves trifoliolate; leaflets more or less elliptic, 8–13 cm long, 4–6 cm wide, glabrous above and below, sometimes weakly barbulate below in some vein axils; tip acuminate; margins serrate-dentate; stipules linear, to 3 mm long; petioles not winged. Tendrils bifid, axillary. Inflorescence a racemiform thyrse, 5.5–20 cm long, axillary, puberulent to pilose. Flowers with pedicels 3.5–4 mm long. Fruit red, to 3 cm long, 3-winged, glabrous.

Phenology. Flowers in November and December. Fruits from January through June.

Distribution. Endemic to Costa Rica. Found from 450 to 1,500 m.

3. *Paullinia bracteosa* Radlk., Bull. Herb. Boissier 2: 321. 1905. Figure 11.1D.

Synonymy. *Paullinia naiguatensis* Steyerm., Bol. Soc. Venez. Ci. Nat. 26: 457. 1966.

Description. Liana; stems 3–5 angled, pubescent to glabrate; wood cross-section with a single vascular cylinder. Leaves 2–3 jugate; leaflets elliptic to obovate, 8–27 cm long, 4–12 cm wide, glabrous above except for hirsute midrib, hispidulous below; tip acuminate; margins crenate above the middle; stipules lanceolate, (2.5) 3–4 cm long, to 1.5 cm wide; petioles and rachis winged, wings to 1.5 cm wide. Tendrils bifid, axillary. Inflorescence axillary, to 20 cm long, rufous tomentose. Sepals tomentulose, to 5.2 mm long; petals white. Fruit pyriform, to 5 cm long, not winged, puberulent to glabrous; stipe to 1.2 cm long. Seeds 1–3, to 2 cm long.

Phenology. Flowers and fruits throughout the year.

Distribution. Nicaragua to Bolivia. In Costa Rica found from sea level to 1,300 m.

4. *Paullinia brenesii* Croat, Ann. Missouri Bot. Gard. 63: 464. 1976. Figure 11.2A.

Description. Liana; stems terete, tomentose to glabrate; wood cross-section with a single vascular cylinder. Leaves biternate; leaflets elliptic-ovate to ovate, 2–10 cm long, 1–5 cm wide, pubescent above and below; tip acuminate or acute; margins crenate to crenate-lobate; stipules linear, to 4 mm long; petiole and rachis not winged or with narrow wings to 3 mm wide. Tendrils bifid, axillary. Inflorescence a terminal thyrse, brown-tomentose. Sepals to 2.7 mm long, pubescent internally; petals white. Fruit red, to 1.2 cm long and 2.5 cm wide, obovate, winged; stipe to 3 mm long. Seeds 1–3.

Phenology. Flowers in September. Fruits in November.

Distribution. Honduras to Panama. In Costa Rica found from 650 to 1,700 m.

5. *Paullinia costaricensis* Radlk., Ergaenz. Monogr. Serjania 157. 1886. Figure 11.2B.

Description. Liana; stems terete, pubescent to glabrate; wood cross-section with a single vascular cylinder. Leaves biternate; leaflets sessile, 5–12 cm long, 1.5–5 cm wide, elliptic-lanceolate to somewhat rhombic, glabrate above, pilose to glabrate below, barbate below in axils of veins; tip narrowly rounded to subacute; base obtuse to attenuate; margins crenate to crenate-dentate or crenate-lobate; stipules minute; petiole not winged, rachis winged. Tendrils bifid, axillary. Inflorescence a solitary thyrse, 8–15 cm long, frequently puberulent. Sepals to 2 mm long; petals white. Fruit red to red-orange, 1–1.2 cm wide, globose, not winged; stipe to 7 mm long. Seeds black, to 7 mm long, with a white arillode.

Phenology. Flowers from July through November and from January through March. Fruits from January through April, and in September and November.

Distribution. Guatemala to Panama. In Costa Rica found from near sea level to 1,400 m.

6. *Paullinia cururu* L., Sp. pl. 365: 1753. Figure 11.2C.

Description. Liana; stems terete to somewhat 6-ribbed, glabrous; wood cross-section with a single vascular cylinder. Leaves trifoliolate; leaflets subsessile, 7–10 cm long, 2.5–5.5 cm wide, elliptic to oblong or ovate, glabrous above, frequently barbate below in the axils of veins; tip acuminate; base acute to obtuse; margins serrate or dentate; stipules more or less triangular, to 4 mm long and 1 mm wide, caducous; petiole winged. Tendrils bifid, axillary. Inflorescence usually a solitary thyrse, 1.7–3.5 cm long, puberulent to glabrate. Flowers with pedicels to 3 mm long; sepals to 4 mm long; petals white. Fruit red, to 2.5 cm long, 0.8–1.2 cm wide, more or less pear- or club-shaped, not winged, glabrous; stipe to 5 mm long. Seeds black, to 1 cm long, with a white arillode.

Phenology. Flowers from April through June. Fruits from June through December.

Distribution. Mexico to Colombia, Venezuela and Brazil. In Costa Rica found from sea level to 1,300 m.

7. *Paullinia faginea* (Triana & Planch.) Radlk., Monogr. Serjania 76, no. 142. 1875. Figure 11.2D.

Synonymy. *Enourea faginea* Triana & Planch., Ann. Sci. Nat., Bot. Sér. 4, 18: 379. 1862; *Paullinia elongata* Radlk., Monogr. Paullinia 238. 1895; *P. florenciana* Cuatrec., Revista Acad. Colomb. Ci. Exact. 4: 347. 1941.

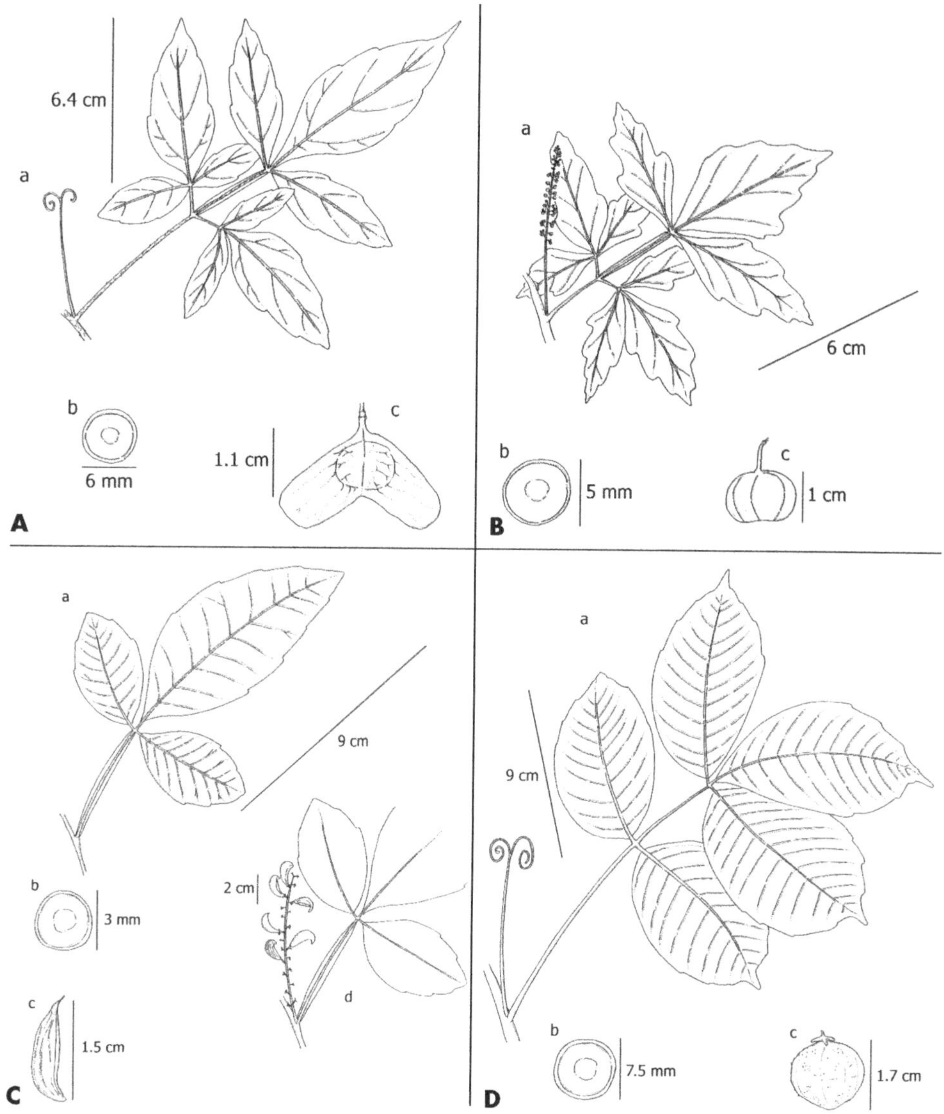

Fig. 11.2. A. *Paullinia brenesii:* **(a)** leaf and tendril, **(b)** stem cross-section showing a single vascular cylinder, **(c)** fruit (Leaf and tendril based on Valerio 81, CR; Stem cross-section and fruit based on Chacón 1568, MO). **B.** *Paullinia costaricensis:* **(a)** leaf and inflorescence, **(b)** stem cross-section showing a single vascular cylinder, **(c)** fruit (Leaf and inflorescence based on Breedlove 30569, US; Stem cross-section and fruit based on Morales 1339, MO). **C.** *Paullinia cururu:* **(a)** leaf, **(b)** stem cross-section showing a single vascular cylinder, **(c)** fruit, **(d)** leaf and infructescence (Leaf and infructescence based on Quigley 904, MO; Stem cross-section and fruit based on Dickson 1328, US). **D.** *Paullinia faginea:* **(a)** leaf and tendril, **(b)** stem cross-section showing a single vascular cylinder, **(c)** fruit (Based on Haber 9302, MO).

Description. Liana; stems terete, rufous-tomentose when young, glabrate in age; wood cross-section with a single vascular cylinder. Leaves pinnately 5-foliolate; leaflets obovate or ovate to elliptic, 6–18 cm long, 3–7 cm wide, pubescent on veins, otherwise glabrate above and below; tip acuminate; margins crenate-serrate; stipules to 10 mm long, elliptic-lanceolate; petioles and rachis not winged. Tendrils bifid, axillary. Inflorescence a racemiform thyrse, 5.5–12 cm long, axillary. Flowers white; sepals rufous-tomentose, to 3 mm long. Fruit globose, to 2.5 cm long, pubescent, not winged.

Phenology. Flowers from June through August. Fruits in March and from June through September.

Distribution. Costa Rica to Peru. In Costa Rica found from 200 to 1,400 m.

8. *Paullinia fuscescens* Kunth in H.B.K., Nov. Gen. sp. pl. 5: 93. 1821. Figure 11.3A.

Synonymy. *Paullinia velutina* DC., Prodr. 1: 605. 1824; *Serjania pubescens* Seem., Bot. voy. Herald 92. 1853, non H.B.K., 1821; *Paullinia fusca* Griseb., Bonplandia 6: 3. 1858.

Description. Liana; stems terete, pilose; wood cross-section with a single vascular cylinder. Leaves biternate; leaflets obovate or ovate to elliptic, 2.5–9 cm long, 1.3–4 cm wide; pubescent above and below (especially along the veins); tip acuminate or rounded; margins crenate; stipules minute; petioles ribbed above; rachis winged. Tendrils bifid, axillary. Inflorescence axillary, 3–10 cm long, pubescent. Sepals densely tomentose, to 2 mm long; petals white. Fruit subglobose to obovoid, to 2.2 cm long, pilose, 3-winged. Seeds 1–3, to 5 mm long, globose.

Phenology. Flowers in January, February, May, September, and November. Fruits in January and May.

Distribution. West Indies, Mexico to Brazil. In Costa Rica found from near sea level to 1,600 m.

9. *Paullinia grandifolia* Benth. ex Radlk., Monogr. Paullinia 227. 1895. Figure 11.3B.

Description. Liana; stems terete, densely tomentulose to glabrate; wood cross-section with a single vascular cylinder. Leaves 3-jugate; lowest pair of leaflets trifoliolate; leaflets oblong-elliptic or lanceolate, 5–14 cm long, 2–5 cm wide, glabrous to somewhat pubescent above and below, not barbate below in axils of veins; tip acuminate; margins crenate; stipules narrowly triangular, to 8 mm long; petioles and rachis slightly winged or not. Tendrils bifid, axillary. Inflorescence axillary or terminal, to 8 cm long, tomentose to hispid. Sepals glabrous internally, pubescent externally, to 4.5 mm long; petals white. Fruit reddish or orange, to 2.3 cm long, subglobose, stipitate, glabrous internally and externally, not winged. Seeds ellipsoid, to 2 cm long.

Phenology. Flowers in April, September, and November. Fruits in February, April through June, and September.

Distribution. Costa Rica to Surinam and Peru. In Costa Rica found from sea level to 1,000 m.

10. *Paullinia ingifolia* Rich. ex Juss., Ann. Mus. Natl. Hist. Nat. 4: 349. 1804. Figure 11.3C.

Description. Liana; stems terete, puberulent or glabrous; wood cross-section with a single vascular cylinder. Leaves 3-jugate, lowest pair of leaflets trifoliolate; leaflets oblong-elliptic or lanceolate, 5–14 cm long, 2–5 cm wide, glabrous to somewhat pubescent above and below, not barbate below in axils of veins; tip acuminate; margins crenate; stipules ovate

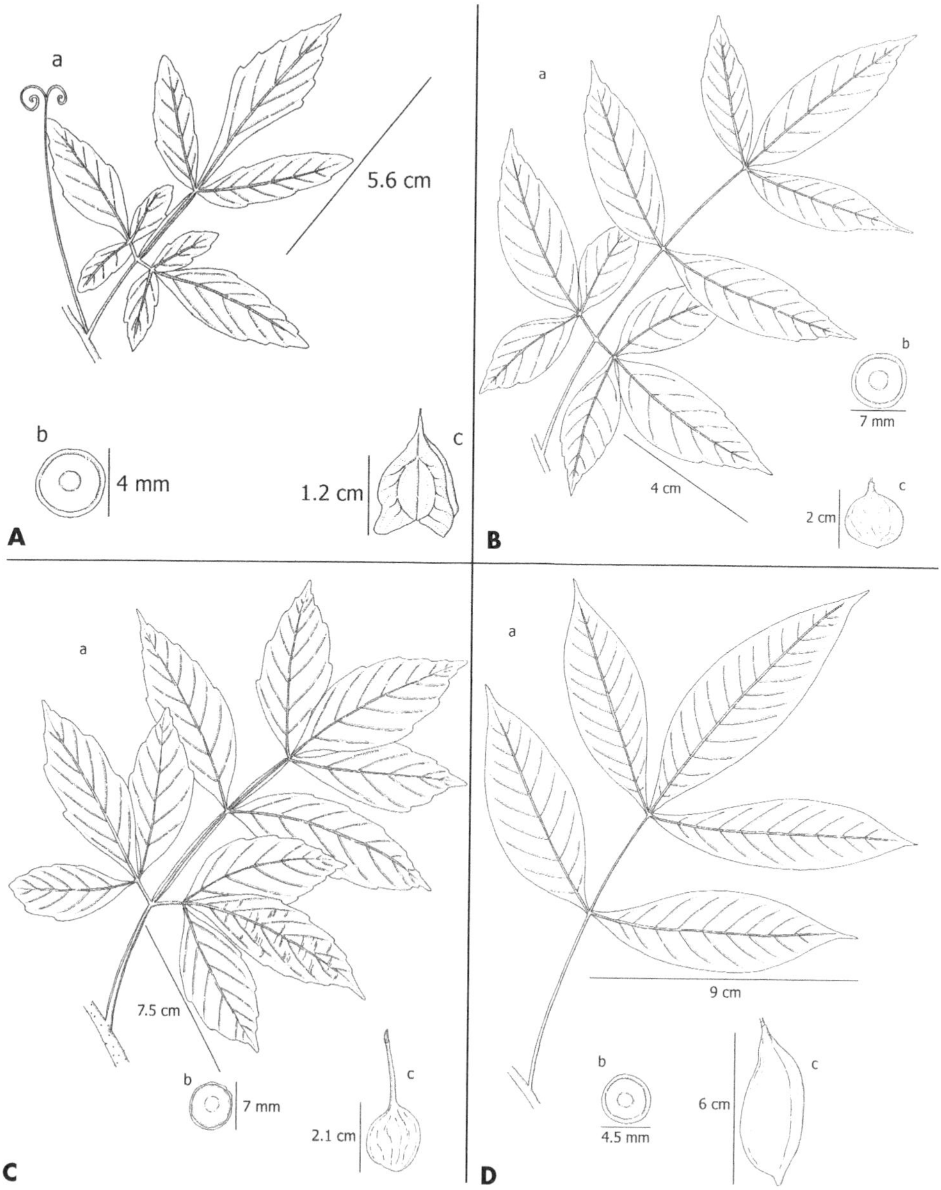

Fig. 11.3. A. *Paullinia fuscescens:* **(a)** leaf and tendril, **(b)** stem cross-section showing a single vascular cylinder, **(c)** fruit (Leaf and tendril based on Espinoza 172, US; Stem cross-section and fruit based on Curran and Haman 780, US). **B.** *Paullinia grandifolia:* **(a)** leaf, **(b)** stem cross-section showing a single vascular cylinder, **(c)** fruit (Leaf based on Moraga 291, US; Stem cross-section and fruit based on Pitman 602, US). **C.** *Paullinia ingifolia:* **(a)** leaf, **(b)** stem cross-section showing a single vascular cylinder, **(c)** fruit (Based on Marin 743, US). **D.** *Paullinia macrocarpa:* **(a)** leaf, **(b)** stem cross-section showing a single vascular cylinder, **(c)** fruit (Leaf based on Haber 7467, US; Stem cross-section and fruit based on Haber ex Bello 6368, MO).

to broadly lanceolate, 1.5–4 cm long, 0.6–0.8 cm wide; petioles slightly winged; rachis winged. Tendrils bifid, axillary. Inflorescence axillary or terminal, to 14 cm long, pubescent. Sepals pubescent externally, to 4.5 mm long; petals white. Fruit subglobose, to 2 cm long, not winged, stipitate; stipe to 1.7 cm long.

Phenology. Flowering and fruiting times unknown in Costa Rica.

Distribution. Largely a South American species. Costa Rica represents the northernmost limit of its distribution.

Notes. This species is included here based on one collection from 740 m in Costa Rica (Boyle and Ezeta 2785, CR).

11. *Paullinia macrocarpa* Radlk., Monogr. Paullinia 260. 1895. Figure 11.3D.

Description. Liana; stems 5-angled, 5-sulcate or terete, glabrous; wood cross-section with a single vascular cylinder. Leaves pinnately 5-foliolate; leaflets oblong or narrowly elliptic, to 14 cm long and 9 cm wide, glabrous above and below; tip acuminate or acute; margins entire to crenate; stipules caducous; petioles and rachis not winged. Tendrils bifid, axillary. Inflorescence axillary, to 10 cm long. Flowers not seen. Fruit red, to 7 cm long, smooth, not winged; stipe to 2 cm long. Seeds 2–3, to 1.8 cm long and 0.8 cm wide.

Phenology. Flowering times unknown. Fruits in September and November.

Distribution. Mexico to Costa Rica. In Costa Rica found at about 900 m.

12. *Paullinia mallophylla* Radlk., Monogr. Paullinia 188: 1895. Figure 11.4A.

Synonymy. *Paullinia verecunda* Standl., Ann. Missouri Bot. Gard. 27: 316. 1940.

Description. Liana; stems 3–5-angled, glabrous, except for the puberulent angles; wood cross-section with a single vascular cylinder. Leaves 3-jugate; lowest pair of leaflets usually trifoliolate; leaflets elliptic, oblong-lanceolate or somewhat rhombic, 2–10 cm long, 1–5 cm wide, pubescent above and below on veins, not distinctly barbate below in axils of veins; tip acuminate or acute; margins entire to serrate-dentate; stipules linear-lanceolate, to 11 mm long, deciduous; petioles not winged; rachis slightly to narrowly winged. Tendrils bifid, axillary. Inflorescence glomerate thyrse, to 2 cm long, axillary, sessile, sparsely villous. Sepals pubescent, to 2 mm long; petals white. Fruit red to brownish, to 1.5 cm long, ellipsoid or globose, pubescent, 3-winged. Seeds subglobose, to 6 mm long.

Phenology. Flowering times unknown. Fruits in January and February.

Distribution. Costa Rica and Panama. In Costa Rica found from near sea level to 700 m.

13. *Paullinia pterocarpa* Triana & Planch., Ann. Sci. Nat., Bot. Sér. 4, 18: 356. 1862. Figure 11.4B.

Synonymy. *Paullinia triptera* Triana & Planch., Ann. Sci. Nat., Bot. Sér. 4, 18: 356. 1862; *P. excisa* Radlk., Monogr. Paullinia 273. 1895; *P. trisulca* Radlk., Bull. Herb. Boissier, Sér. 2, 5: 323. 1905; *P. vaupesana* Killip & Cuatrec., Revista Acad. Colomb. Ci. Exact. 4: 344. 1941.

Description. Liana; stems terete to 3-ribbed, glabrous; wood cross-section with a single vascular cylinder. Leaves pinnately 5-foliolate; leaflets ovate to elliptic, 5–19 cm long, 2.2–9 cm wide, glabrous above and below; tip acuminate or acute; margins entire to serrate-dentate; stipules linear-lanceolate, to 12 mm long; petioles and rachis not winged. Tendrils bifid, axillary. Inflorescence glomerulate (rarely solitary), 3–5.5 cm long, axillary, puberulous to glabrate. Flowers white; sepals puberulent to glabrous, to 2.5 mm long. Fruit red, to 3.5 cm long, ovoid, glabrous, strongly 3-winged.

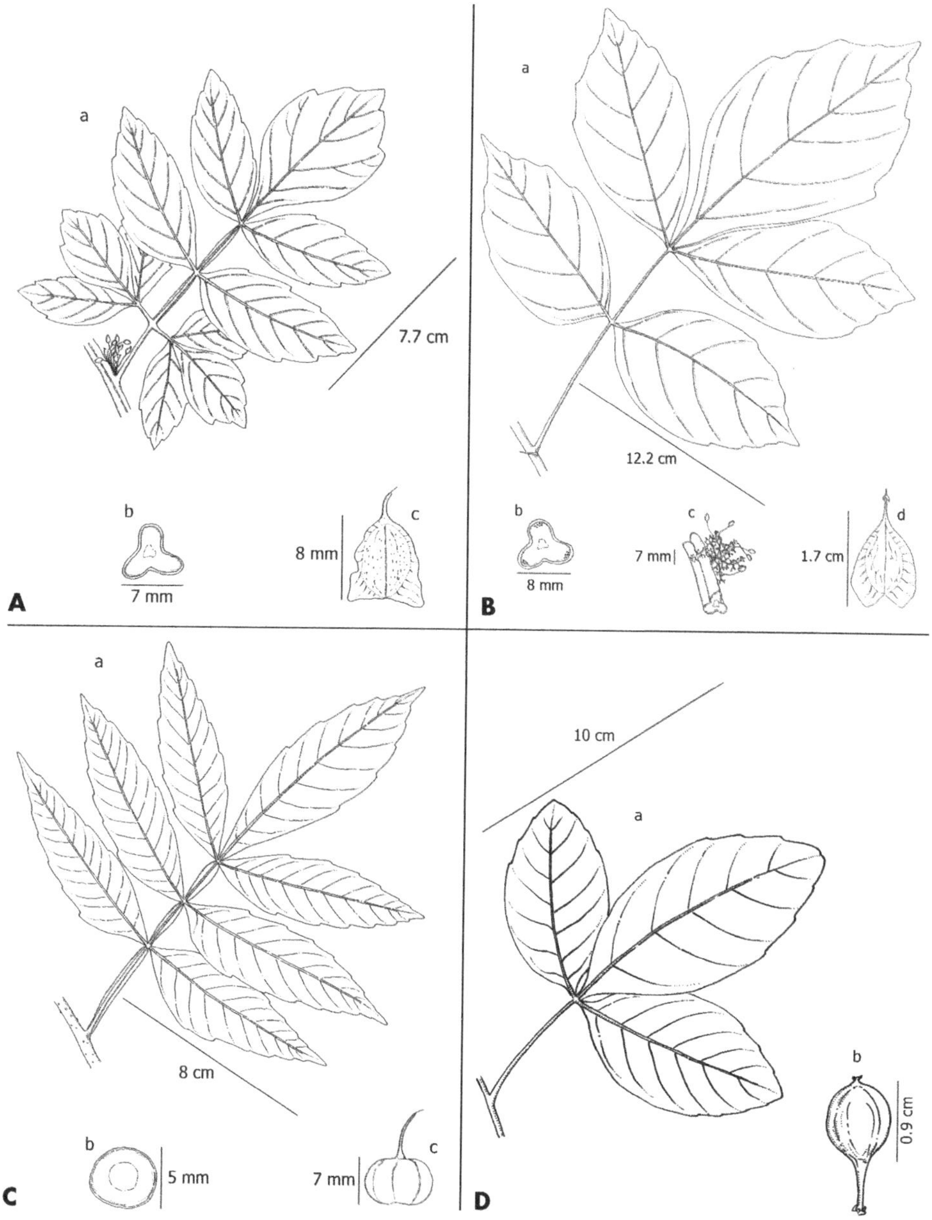

Fig. 11.4. A. *Paullinia mallophylla:* **(a)** leaf and inflorescence, **(b)** stem cross-section showing a single vascular cylinder, **(c)** fruit (Leaf and inflorescence based on Nee 8952, DUKE; Stem cross-section and fruit based on Standley and Valerio 46166, US).
B. *Paullinia pterocarpa:* **(a)** leaf, **(b)** stem cross-section showing a single vascular cylinder, **(c)** inflorescence, **(d)** fruit (Leaf based on Hammel and Trainer 10876, DUKE; Inflorescence based on Hammel 9577, DUKE; Stem cross-section and fruit based on Allen 5791, US).
C. *Paullinia pterophylla:* **(a)** leaf, **(b)** stem cross-section showing a single vascular cylinder, **(c)** fruit (Leaf based on Moreno 23696, DUKE; Stem cross-section and fruit based on Ríos 357, US). **D**. *Paullinia talamancensis:* **(a)** leaf, **(b)** fruit (Based on Davidse et al. 25563, MO).

Phenology. Flowers in September. Fruits in January, February and from May through November.
Distribution. Costa Rica to Peru. In Costa Rica found from 500 to 1,100 m.

14. *Paullinia pterophylla* Triana & Planch., Ann. Sci. Nat., Bot. Sér. 4, 18: 354. 1862. Figure 11.4C.

Description. Liana; stems subterete to sulcate, scattered pubescent to glabrescent; wood cross-section with a single vascular cylinder. Leaves pinnately 7–9-foliolate; leaflets elliptic to lanceolate-oblong, to 9 cm long and 4 cm wide, glabrous above and below; tip acuminate; margins crenate to serrate-dentate; stipules small, subulate; petioles and rachis winged. Tendrils bifid, axillary. Inflorescence in axillary thyrse, to 6 cm long, pubescent to glabrate. Flowers with petals white. Fruit somewhat 3-angled, to 1.1 cm long and 1.8 cm wide, not winged.
Phenology. Flowering times unknown. Fruits in April.
Distribution. Nicaragua to Ecuador. In Costa Rica found from 100 to 700 m.

15. *Paullinia talamancensis* J.F. Morales, Phytologia 81: 363–364. 1996. Figure 11.4D.

Description. Liana; stems subcylindrical, glabrate; cross-section with a single vascular cylinder. Leaves trifoliolate; leaflets 1.5–4.5 cm long, 3–5 cm wide, elliptic-oblong, glabrate, subsessile to petiolulate; tip obtuse or acute; base rounded; margins entire to serrate; petiole not winged; stipules scarious. Tendrils bifid, axillary. Inflorescence racemose, solitary, axillary, 8–15 cm long, glabrate. Flowers unknown. Fruit red, 0.9–1.2 cm long, suborbicular, stipitate, not winged.
Phenology. Flowering times unknown. Fruits in March.
Distribution. Probably endemic to Costa Rica, found from 1,150 to 1,250 m, but possibly also found in Panama (Morales 2000).

3. GENUS *SERJANIA*

Vegetatively *Serjania* closely resembles *Paullinia*. *Serjania* is distinguished from all other genera in the Sapindaceae by the schizocarp fruit that splits into three proximally winged mericarps. Most of our *Serjania* are biternate, and only two (*S. valerii* and *S. lobulata*) are consistently pinnate. *Serjania* with trifoliolate leaves is principally restricted to elevations below 700 m. *Serjania mexicana* frequently has prickles on the stems and is the only armed sapindaceous species in our area.

KEY TO *SERJANIA*

1. Leaves pinnately 5-foliolate
    2. Stems and petioles hirtellous; hairs 0.1–0.3 mm long; stipules triangular, to 3 mm long and 1 mm wide; fruit to 2 cm long... *S. lobulata*, p. 145.
    2. Stems and petioles densely hirsute; hairs to 1–2 mm long; stipules lacerate or fringed, to 5 mm long and 7 mm wide; fruit to 4 cm long... *S. valerii*, p. 150.
1. Leaves biternate
    3. Stems armed with prickles or thorns... *S. mexicana*, p. 146.
    3. Stems unarmed
        4. Wood cross-section with a single vascular cylinder
            5. Seed-bearing portion of fruit pubescent

6. Leaflets elongate, narrowly elliptic to narrowly lanceolate; plant rufous-villous to rufous-pubescent... *S. acuta*, p. 143.
6. Leaflets ovate to broadly elliptic; plant rufous-pubescent or not... *S. rufisepala*, p. 150.

5. Seed-bearing portion of fruit glabrous
7. Leaflet margins crenate to crenate-lobate; fruit 2.7–3.5 cm long, 2–2.6 cm wide; seed-bearing portion 1/6 to 1/3 the length of the fruit... *S. membranacea*, p. 145.
7. Leaflet margins serrate to serrate-lobate; fruit 1.5–2.4 (2.8) cm long, 1–1.8 cm wide; seed-bearing portion 1/3 to nearly 1/2 the length of the fruit... *S. racemosa*, p. 148.

4. Wood cross-section with 3–10 peripheral vascular bundles surrounding the central vascular bundle
8. Wood cross-section with 8–10 (or more) peripheral vascular cylinders, arranged in a circle around the central cylinder... *S. caracasana*, p. 145.
8. Wood with 3–5 peripheral vascular cylinders, arranged in a triangle around the central cylinder (sometimes 1–2 of the peripheral vascular cylinders further split into two, but the triangular arrangement remains)
9. Leaflets black-lineolate below (at least when dry); sepals conspicuously black-lineolate... *S. atrolineata*, p. 143.
9. Leaves not black-lineolate below; sepals not lineolate or obscurely so
10. Leaves densely pubescent to pilose below... *S. rhombea*, p. 148.
10. Leaves glabrous below (sometimes with veins slightly strigillose or vein axils barbulate)
11. Rachis not winged; wood with five peripheral vascular cylinders surrounding the central vascular cylinder... *S. paniculata*, p. 146.
11. Rachis winged; wood with three peripheral vascular cylinders surrounding the central vascular cylinder... *S. paucidentata*, p. 148.

1. *Serjania acuta* Triana & Planch., Ann. Sci. Nat., Bot. Sér. 4, 18: 349. 1862. Figure 11.5A.

Synonymy. *Serjania seemannii* Triana & Planch., Ann. Sci. Nat., Bot. Sér. 4, 18: 346. 1862; *S. salzmanniana* Seem., Bot. voy. Herald 92. 1857, non Schltdl. 1844; *S. sordida* Radlk., Monogr. Serjania 272. 1875.

Description. Liana; stems terete to 6-costate, rufous-villous to rufous-pubescent; wood cross-section with a single vascular cylinder. Leaves biternate; leaflets elliptic to elliptic-lanceolate, 2–11 cm long, 1–5 cm wide, glabrous to pubescent above, rufous-villous to tomentose below; tip acuminate; base obtuse to attenuate; margins crenate to serrate; petioles and rachis not winged. Tendrils bifid, axillary. Inflorescence a thyrse, 5–10 cm long, axillary, frequently rufous-tomentose. Flowers to 4 mm long; sepals rufous-tomentose; petals white. Fruit to 2.5 cm long and 2 cm wide; seed-bearing portion puberulent.

Phenology. Flowers in January and March. Fruits in March.

Distribution. Mexico to Venezuela. In Costa Rica found from sea level to 1,200 m.

2. *Serjania atrolineata* Sauv. & Wright, Fl. cub. 24. 1873. Figure 11.5B.

Synonymy. *Serjania scatens* Radlk., Monogr. Serjania 213. 1875; *S. atrolineata* f. *lancifolia* Radlk., Repert. Spec. Nov. Regni Veg. 17: 357. 1921.

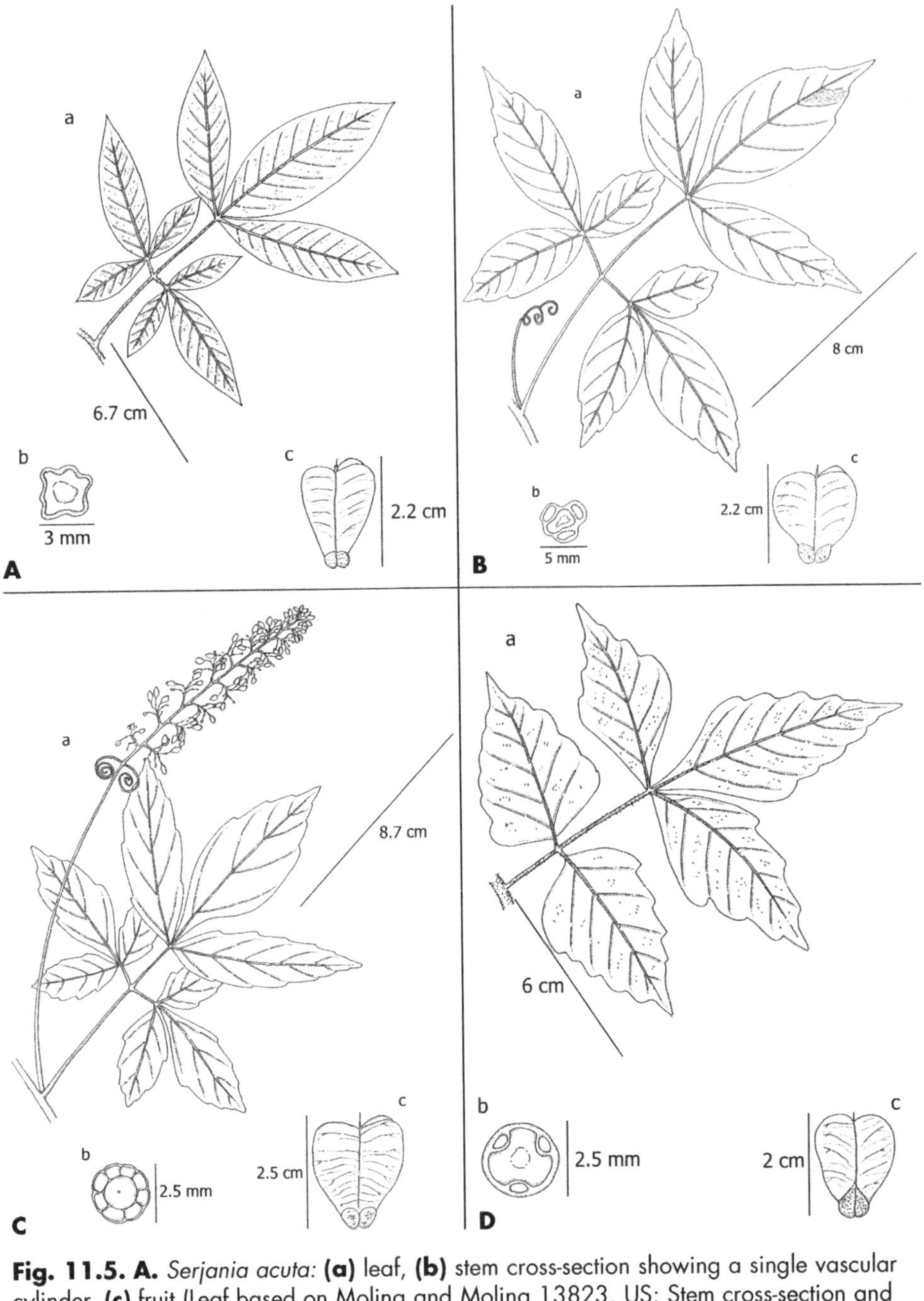

Fig. 11.5. A. *Serjania acuta:* **(a)** leaf, **(b)** stem cross-section showing a single vascular cylinder, **(c)** fruit (Leaf based on Molina and Molina 13823, US; Stem cross-section and fruit based on Standley 49357, US). **B.** *Serjania atrolineata:* **(a)** leaf and tendril, **(b)** stem cross-section, **(c)** fruit (Leaf based on Bunting and Licht 640, US; Stem cross-section based on Tonduz 9360, US; Fruit based on Conzatti 1010, US). **C.** *Serjania caracasana:* **(a)** leaf and inflorescence, **(b)** stem cross-section showing eight peripheral vascular cylinders surrounding a central vascular cylinder, **(c)** fruit (Leaf and inflorescence based on Liesner and Dwyer 1587, DUKE; Stem cross-section and fruit based on Espinoza 202, US). **D.** *Serjania lobulata:* **(a)** leaf, **(b)** stem cross-section showing three peripheral vascular cylinders surrounding a central vascular cylinder, **(c)** fruit (Based on Acevedo-Rodríguez 451, MO).

Description. Liana; stems ribbed, glabrous; wood cross-section with three peripheral vascular cylinders surrounding the central vascular cylinder in a more or less triangular arrangement; sap often drying blackish, somewhat resinous. Leaves biternate; leaflets lanceolate, elliptic-lanceolate or obovate, 5–9 cm long, 1.3–3.3 cm wide, glabrous above and below (sometimes pubescent below on veins), frequently black-lineolate below; tip acuminate to obtuse; margins crenate to serrate; petioles and rachis margined to narrowly winged. Tendrils bifid, axillary. Inflorescence a thyrse, 4–10 cm long, axillary, glabrous to scattered pubescent. Flowers to 4 mm long; sepals densely tomentose, lineolate; petals white. Fruit 2.5 cm long and 2.2 cm wide; seed-bearing portion generally pubescent.

Phenology. Flowers in October and from January through March. Fruits in September and from February through April.

Distribution. Cuba and Mexico to Venezuela. In Costa Rica found from near sea level to 1,100 m.

3. *Serjania caracasana* (Jacq.) Willd., Sp. pl. 2: 465. 1799. Figure 11.5C

Synonymy. *Paullinia caracasana* Jacq., Hort. Schoenbr. 1: 52, pl. 99. 1797; *P. glabra* Bertol., Fl. guatimal. 413, pl. 40. 1840.

Description. Liana; stems terete or shallowly 6–8 ribbed, glabrous; wood cross-section with 8–10 peripheral vascular cylinders in a circle surrounding the central vascular cylinder. Leaves biternate; leaflets narrowly elliptic to lanceolate, 5.5–11 cm long, 3–5 cm wide, glabrous above and below; tip acuminate to obtuse; margins crenate to serrate; petioles and rachis not winged. Tendrils bifid, axillary. Inflorescence in axillary thyrse, 5–24 cm long, frequently puberulent. Flowers to 6 mm long; pedicels to 6 mm long; sepals glabrous to puberulent; petals white. Fruit 2.5–4 cm long, 2.5–4 cm wide; seed-bearing portion glabrous; wings not constricted below the seed-bearing portion.

Phenology. Flowers in January and June. Fruits in February.

Distribution. Mexico to Paraguay. In Costa Rica found from near sea level to 1,000 m.

4. *Serjania lobulata* Standl. & Steyerm., Publ. Field Mus. Nat. Hist., Bot. Ser. 23: 14. 1943. Figure 11.5D.

Description. Liana; stems terete to somewhat angled, hirtellous; stem hairs erect, 0.1–0.3 mm long; wood cross-section with 3 peripheral vascular cylinders surrounding the central vascular cylinder. Leaves pinnately 5-foliolate; leaflets ovate to elliptic, 3–9 cm long, 1.5–6 cm wide, hirtellous above and below; tip acuminate; margins crenate-lobate or dentate-lobate; stipules triangular, to 3 mm long and 1 mm wide; petioles and rachis not winged. Tendrils bifid, axillary. Inflorescence a thyrse, 3–12 cm long, solitary, axillary or terminal. Flowers to 4.5 mm long; sepals pubescent; petals white. Fruit to 2 cm long and 1.9 cm wide; seed-bearing portion velutinous.

Phenology. Flowers in August. Fruiting times unknown.

Distribution. Mexico to Costa Rica. In Costa Rica found from 1,400 to 1,500 m.

5. *Serjania membranacea* Splitg., Tijdschr. Natuurl. Gesch. Physiol. 9: 105. 1842. Figure 11.6A.

Description. Liana; stems terete or sometimes sulcate, pubescent to glabrous, frequently hollow when dry; wood cross-section with a single vascular cylinder. Leaves biternate; leaflets elliptic or ovate, 3–9 cm long, 2–5 cm wide, pubescent along the veins and sometimes above and below; tip acuminate to obtuse; margins crenate to crenate-lobate; petiole not winged; rachis slightly winged. Tendrils bifid, axillary. Inflorescence in axillary thyrse or terminal paniclelike thyrsoids, 4–6 cm long, pilose. Flowers to 4 mm long; sepals with minute, white hairs and ciliate margins; petals white. Fruit 2.7–3.5 cm long, 2–2.6 cm wide; seed-bearing portion glabrous.

Phenology. Flowers in October. Fruiting times unknown.

Distribution. Largely South American. Costa Rica is the northern-most limit of its distribution. In Costa Rica found from 800 to 900 m.

6. *Serjania mexicana* (L.) Willd., Sp. pl. 2: 465. 1799. Figure 11.6B.

Synonymy. *Paullinia mexicana* L., Sp. pl. 366. 1753.

Description. Liana; stems 5-angled, sulcate, with small prickles or stout thorns to 2.5 cm long, pubescent to glabrous; wood cross-section with a single vascular cylinder. Leaves biternate; leaflets ovate to elliptic-ovate, 5–13 cm long, 1.7–7 cm wide, glabrous above, glabrous to somewhat puberulent below; tip acute to obtuse; margins entire to crenate-serrate above middle; petioles not winged; rachis narrowly to broadly winged. Tendrils bifid, axillary. Inflorescence a thyrse, 4.5–10 cm long, axillary, frequently puberulent. Flowers to 4 mm long; sepals tomentulose; petals white. Fruit to 2.8 cm long and 2 cm wide; seed-bearing portion flattened, glabrous.

Phenology. Flowers from December through April. Fruits in March and April.

Distribution. Mexico to Colombia. In Costa Rica found from sea level to 1,300 m.

7. *Serjania paniculata* Kunth in H.B.K., Nov. Gen. sp. pl. 5: 111. 1821. Figure 11.6C.

Description. Liana; stems about 5-ribbed, tomentulose, puberulent or glabrous; wood cross-section with five peripheral vascular cylinders surrounding the central cylinder; peripheral vascular cylinders sometimes split into two. Leaves biternate; leaflets narrowly elliptic to obovate or ovate, 3–11 cm long, 2–5 cm wide, glabrous but strigillose on veins, barbulate below in axils of veins; tip acuminate to obtuse; margins crenate to serrate or crenate-lobate; petiole not winged; rachis sometimes slightly winged. Tendrils bifid, axillary. Inflorescence a thyrse, 4–8 cm long, axillary or terminal, frequently tomentulose or villous. Flowers to 4 mm long; sepals tomentose; petals white. Fruit 2–2.8 cm long, 1.5–1.9 cm wide; seed-bearing portion whitish-pubescent.

Phenology. Flowers in December. Fruits in March.

Distribution. Mexico to Colombia and Venezuela. In Costa Rica found from sea level to 850 m.

Fig. 11.6. A. *Serjania membranacea:* **(a)** leaf and inflorescence, **(b)** stem cross-section showing a single vascular cylinder, **(c)** fruit (Leaf, inflorescence, and stem cross-section based on Herréra 815, US; Fruit based on Ferreira 8677, US). **B.** *Serjania mexicana:* **(a)** leaf, **(b)** stem cross-section showing a single vascular cylinder, **(c)** fruit (Leaf based on Grayum 3076, DUKE; Stem cross-section and fruit based on Tonduz 9614, US). **C.** *Serjania paniculata:* **(a)** leaf, **(b)** stem cross-section showing five peripheral vascular cylinders surrounding a central vascular cylinder, **(c)** stem cross-section showing seven peripheral vascular cylinders surrounding a central vascular cylinder, **(d)** fruit (Leaf based on

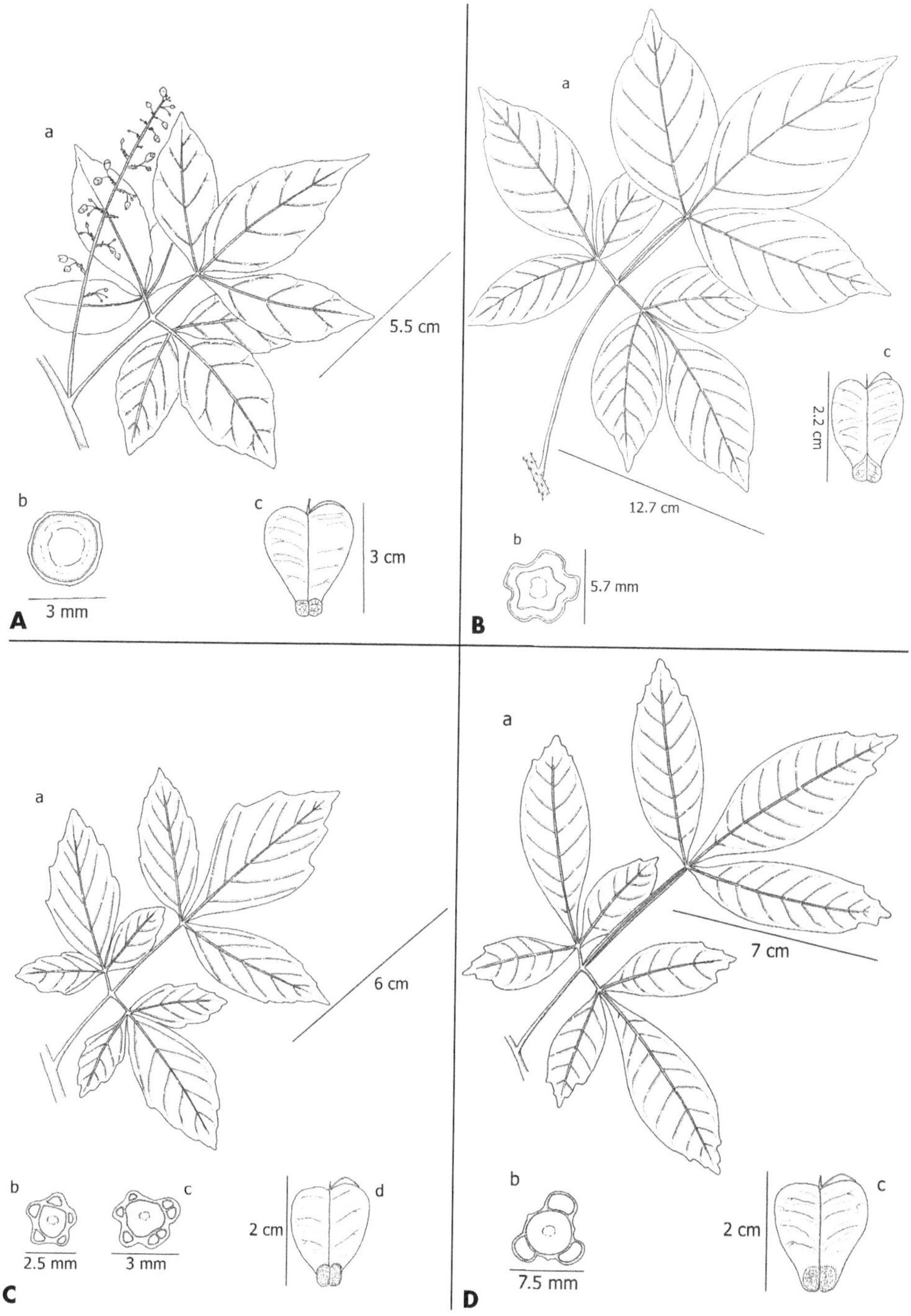

Hinton et al. 11741, US. Fruit and stem cross-sections Walker 1197, US; and Kellerman 6492, US). **D.** *Serjania paucidentata:* **(a)** leaf, **(b)** stem cross-section showing three peripheral vascular cylinders surrounding a central vascular cylinder, **(c)** fruit (Leaf based on Williams and Molina R. 17971, US; Stem cross-section based on Mexía 9220, US; Fruit based on Skutch 4811, US).

8. *Serjania paucidentata* DC., Prodr. 1: 603. 1824. Figure 11.6D.

Synonymy. *Paullinia protracta* Steud., Flora 27: 725. 1844.

Description. Liana; stems 3–6-ribbed, glabrous; wood cross-section with three peripheral vascular cylinders in a triangle surrounding the central vascular cylinder. Leaves biternate; leaflets narrowly elliptic, 3–12 cm long, 2–5 cm wide, glabrous above and below; tip acuminate to acute, frequently somewhat toothed; margins crenate to serrate above the middle; rachis narrowly winged. Tendrils bifid, axillary. Inflorescence a thyrse, axillary or terminal, frequently tomentose. Flowers to 4.5 mm long; sepals tomentose; petals white. Fruit 2–2.8 cm long, 1.5–1.9 cm wide; seed-bearing portion brownish-pubescent.

Phenology. Flowering times unknown. Fruits in January.

Distribution. Mexico to the Guianas, Trinidad, Peru. In Costa Rica found from sea level to 900 m.

9. *Serjania racemosa* Schumach., Skr. Naturhist.-Selsk. 3, pt. 2: 127, pl. 12, f. 3. 1794. Figure 11.7A.

Description. Liana; stems 5–6-sulcate, pubescent to glabrous; wood cross-section with a single vascular cylinder. Leaves biternate; leaflets elliptic or ovate, 3–9 cm long, 2–5 cm wide, glabrous above and below, sometimes puberulent along the veins; tip acuminate to obtuse; margins serrate to serrate-lobate (rarely crenate); petiole not winged; rachis narrowly winged. Tendrils bifid, axillary. Inflorescence in axillary thyrse or terminal paniclelike thyrsoids, 4–6 cm long, pilose to glabrous. Flowers to 4 mm long; sepals glabrous or with scattered, minute, white hairs and ciliate margins; petals white. Fruit 1.5–2.4 (2.8) cm long, 1–1.8 cm wide; seed-bearing portion glabrous.

Phenology. Flowers from October through March and in July. Fruits from December through February and in July.

Distribution. Mexico to Costa Rica. In Costa Rica found from 600 to 1,700 m.

10. *Serjania rhombea* Radlk., Monogr. Serjania 324. 1875. Figure 11.7B.

Synonymy. *Serjania paucidentata* Seem., Bot. voy. Herald 92. 1853, non DC., 1857; *S. pubescens* Griseb., Bonplandia 6: 3. 1858, non Kunth, 1821, non Seem., 1853; *Paullinia velutina* Triana & Planch., Ann. Sci. Nat., Bot. Sér. 4, 18: 358. 1862; *Serjania microcephala* Radlk., Smithsonian Misc. Collect. 61: 1. 1914.

Description. Liana; stems 5- to 6-ribbed, sometimes somewhat trigonal, brownish hirtellous; wood cross-section with three peripheral vascular cylinders surrounding the central vascular cylinder. Leaves biternate; leaflets ovate or rhombic, 3–10 cm long, 1–5 cm wide, pilosulose above especially on veins, pilose below; tip acuminate to obtuse; margins crenate to serrate usually above middle; rachis narrowly winged. Tendrils bifid, axillary. Inflorescence in axillary thyrses or terminal paniclelike thyrsoids, 4–8 cm long, frequently villous. Flowers to 3 mm long; sepals tomentulose; petals white. Fruit 1.5–2.3 cm long, 1.5–2.2 cm wide; seed-bearing portion pubescent.

Phenology. Flowers from September through January. Fruits from November through February and in August.

Distribution. Mexico to Venezuela and Ecuador. In Costa Rica found from near sea level to 2,000 m.

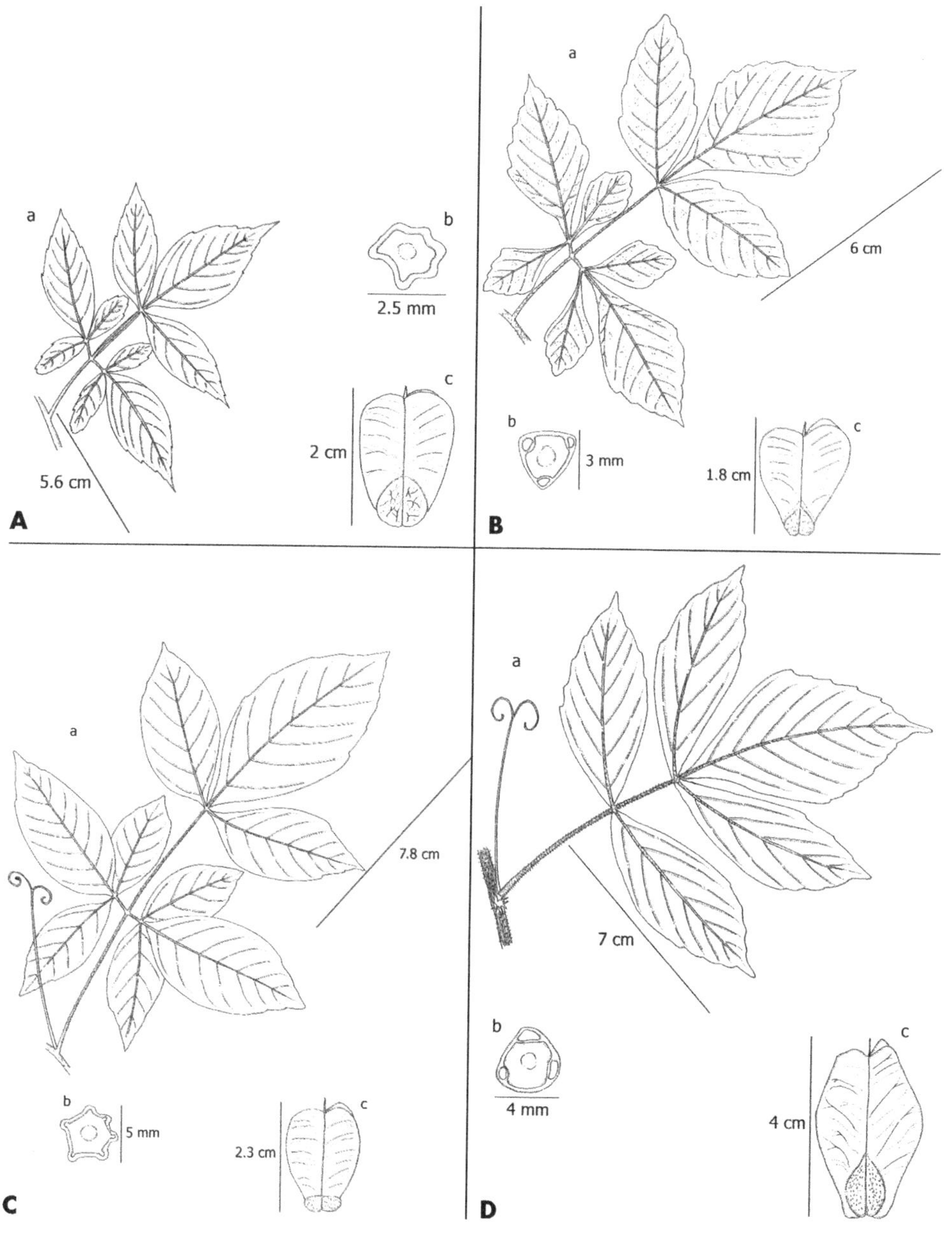

Fig. 11.7. A. *Serjania racemosa:* **(a)** leaf, **(b)** stem cross-section showing a single vascular cylinder, **(c)** fruit (Leaf and stem cross-section based on Skutch 2350, US; Fruit based on Standley 64615, US). **B.** *Serjania rhombea:* **(a)** leaf, **(b)** stem cross-section showing three peripheral vascular cylinders surrounding a central vascular cylinder, **(c)** fruit (Leaf based on Pittier 4771, US; Stem cross-section and fruit based on Wilbur and Stone 9714, US). **C.** *Serjania rufisepala:* **(a)** leaf and tendril, **(b)** stem cross-section showing a single vascular cylinder, **(c)** fruit (Based on Pittier 16368, US). **D.** *Serjania valerii:* **(a)** leaf and tendril, **(b)** stem cross-section showing three peripheral vascular cylinders surrounding a central vascular cylinder, **(c)** fruit (Leaf and tendril based on Krings 154, CR; Stem cross-section and fruit based on Standley and Valerio 45835, US).

11. ***Serjania rufisepala*** Radlk. in Donn. Sm., Bot. Gaz. 16: 191. 1891. Figure 11.7C.

Description. Liana; stems 6-sulcate, glabrous or somewhat rufous-pubescent; wood cross-section with a single vascular cylinder. Leaves biternate; leaflets ovate to broadly elliptic, to 8 cm long and 5 cm wide, usually glabrous above and below, sometimes rufous-pubescent below; tip acuminate; margins crenate to serrate; petiole and rachis not winged. Tendrils bifid, axillary. Inflorescence a thyrse, axillary, frequently puberulent. Flowers to 4 mm long; sepals rufous-tomentose; petals white or yellowish-white. Fruit to 2 cm long and 1.8 cm wide; seed-bearing portion puberulent.

Phenology. Flowering times unknown. Fruits in March.

Distribution. Guatemala to Panama. In Costa Rica found from 1,000 to 1,600 m.

Notes. Vegetatively very similar to *S. acuta*, but leaves somewhat broader.

12. ***Serjania valerii*** Standl., Publ. Field Mus. Nat. Hist., Bot. Ser. 18: 646. 1938. Figure 11.7D.

Description. Liana; stems about 5-angled, densely hirsute; stem hairs more or less erect, 1–2 mm long; wood cross-section with three peripheral vascular cylinders surrounding the central cylinder. Leaves pinnately 5-foliolate; leaflets oblong, elliptic or ovate, 6–13.5 cm long, 3–5 cm wide, hirsute above and below; tip acuminate; margins coarsely dentate; stipules to 5 mm long and 7 mm wide, often lacerate or fringed; petiole and rachis not winged. Tendrils bifid, axillary. Inflorescence a thyrse, axillary, hirtellous. Flowers to 6 mm long; sepals tomentulose; petals white. Fruit to 4 cm long and 2 cm wide; seed-bearing portion flattened, tomentulose.

Phenology. Flowers and fruits in January.

Distribution. Endemic to Costa Rica. Found from 600 to 1,200 m.

Notes. The original description noted that stems have only a single vascular cylinder. Although we did not examine the type specimen (Brenes 4532, F), the cross-section of the other specimen (Standley and Valerio 45835, US) cited in the original description clearly shows three peripheral vascular cylinders surrounding the central vascular cylinder. These three vascular cylinders easily separate *S. valerii* from lowland *Paullinia rugosa,* which has only one vascular cylinder in cross-section.

12
SMILACACEAE

Smilacaceae is a relatively small family of four genera and roughly 375 species found in tropical and temperate regions (Chant 1978). Represented by 13 species in our area, *Smilax* is the only tendrillate monocot. Vegetatively, the genus is easily recognized by paired, stipular tendrils. Two species are endemic to Costa Rica, *S. angustiflora* and *S. candelariae.*

Identification using only stems and leaves is difficult among *Smilax*. For reliable determinations, fruits or, better yet, flowers are required. Since *Smilax* is dioecious, both sexes may be required for identification. Other information on this family is found in Huft (1994).

1. GENUS *SMILAX*

Our treatment follows the revision of the Mexican and Central American species of *Smilax* by Killip and Morton (1936) with three exceptions: (1) addition of *S. chiriquensis* (see Woodson and Schery 1942), (2) recognition of *S. domingensis*, previously considered a synonym of *S. lanceolata* (see Morton 1962), and (3) reduction of *S. standleyi* to synonymy with *S. vaniliodora* following an annotation by Huft on the type specimen (Standley and Valerio 45557, US). By not recognizing *S. domingensis* as synonymous with *S. lanceolate*, we depart from Standley (1937) and Croat (1978), but remain consistent with Morton (1945, 1962).

Based on the difficulty of distinguishing the fruit of *S. domingensis* (*S. lanceolata* sensu Croat [1978]) from *S. spinosa*, Croat (1978) questioned the validity of the latter species. We tentatively separate these two species based on the color of dried leaves. Leaves of *Smilax domingensis* generally become dark reddish-brown to dark olive when dry, whereas those of *S. spinosa* are light green to light olive. In addition, specimens of *S. spinosa* at US sometimes have prickles on leaf undersurfaces near the midrib base, a character not observed in *S. domingensis*.

Our treatment of all species is tentative, acknowledging the need for a revision of the genus. Keys and descriptions draw heavily from Killip and Morton (1936). Two keys are provided, the first using vegetative characters and flowers, and the second using vegetative characters and fruits.

KEY TO *SMILAX* BASED UPON VEGETATIVE CHARACTERS AND FLOWERS

1. Plants never armed
    2. Young branchlets sharply tetragonal, angles often ribbed, glabrate in age; staminodia 6... *S. subpubescens*, p. 158.
    2. Young branchlets terete to subangulate (if somewhat tetragonal, the angles rounded), puberulent to pubescent in age; staminodia 3
        3. Branchlets densely woolly-tomentose or yellowish velutinous... *S. velutina*, p. 160.
        3. Branchlets puberulent, pilose, or subtomentose; hairs coarse, somewhat scattered, not dense
            4. Peduncle of inflorescence shorter than 0.5 cm, shorter than subtending petiole... *S. candelariae*, p. 153.
            4. Peduncle of inflorescence longer than 1.2 cm, longer than subtending petiole

5. Leaves generally 3-plinerved; perianth short-hairy externally; segments 6.5+ mm long, lacking an apical tuft of hair... *S. angustiflora*, p. 153.
5. Leaves generally basally nerved; perianth glabrous externally (sometimes with a few scattered hairs); segments less than 5 mm long, with an apical tuft of long hairs... *S. mollis*, p. 156.

1. Plants frequently armed
   6. Staminate flowers shorter than 2.8 mm... *S. spinosa*, p. 157.
   6. Staminate flowers longer than 4 mm
      7. Peduncles of pistillate inflorescence shorter than subtending petiole, flattened or not
         8. Primary veins distinctly impressed above; secondary veins inconspicuous; pistillate flowers with six staminodia... *S. engleriana*, p. 155.
         8. Primary veins distinctly raised above or very shallowly impressed; secondary veins noticeably reticulate; pistillate flowers with three staminodia
            9. Leaves 5-veined... *S. domingensis*, p. 155.
            9. Leaves 7-veined... *S. kunthii*, p. 155.
      7. Peduncles of pistillate inflorescence longer than subtending petiole, conspicuously flattened
         10. Filaments longer than anthers; stems tetragonal... *S. chiriquensis*, p. 153.
         10. Filaments shorter than anthers; stems tetragonal or not
            11. Stems sharply tetragonal... *S. vanilliodora*, p. 159.
            11. Stems terete to subangulate
               12. Leaves 5–7-veined; secondary veins reticulate; staminate flowers pedicellate, borne on axillary, raceme-like branchlets... *S. panamensis*, p. 157.
               12. Leaves 3–5-veined; secondary veins parallel; staminate flowers sessile, borne in solitary, axillary umbels... *S. spissa*, p. 158.

KEY TO *SMILAX* BASED UPON VEGETATIVE CHARACTERS AND FRUIT

1. Plants unarmed
   2. Branchlets (and often leaves) densely yellowish woolly-tomentose or velutinous... *S. velutina*, p. 160.
   2. Branchlets and leaves variously puberulent to pubescent, but not densely yellowish woolly-tomentose or velutinous
      3. Peduncle of fruiting umbel longer than subtending petiole
         4. Leaf bases broadly attenuate to rounded, not cordate, secondary veins distinctly parallel... *S. spissa*, p. 158.
         4. Leaf bases cordate or subcordate, secondary veins reticulate
            5. Leaves generally 3-plinerved... *S. angustiflora*, p. 153.
            5. Leaves generally basally nerved... *S. mollis*, p. 156.
   2. Peduncle of fruiting umbel shorter than subtending petiole
      6. Young branchlets terete to subangulate (if somewhat tetragonal, angles rounded, but not sharp or ribbed)... *S. candelariae*, p. 153.
      6. Young branchlets sharply tetragonal, angles often ribbed... *S. subpubescens*, p. 158.
1. Plants armed
   7. Peduncle of fruiting umbel shorter than subtending petiole
      8. Primary leaf veins distinctly impressed above, secondary veins inconspicuous above and frequently inconspicuous below... *S. engleriana*, p. 155.
      8. Primary leaf veins raised or only very slightly impressed above, secondary veins conspicuously reticulate, especially below

9. Leaves 7-veined... *S. kunthii*, p. 155.
9. Leaves 5-veined
10. Leaves drying dark reddish-brown, dark green or dark olive, in similar shades above and below, or not... *S. domingensis*, p. 155.
10. Leaves drying light olive to light green, in similar shades above and below... *S. spinosa*, p. 157.
7. Peduncle of fruiting umbel longer than subtending petiole
11. Stems tetragonal
12. Principal leaf veins raised above (at least on dried specimens)... *S. chiriquensis*, p. 153.
12. Principal leaf veins impressed above (at least on dried specimens)... *S. vanilliodora*, p. 159.
11. Stems terete or only subangular
13. Leaves (5-) 7-veined, secondary veins reticulate... *S. panamensis*, p. 157.
13. Leaves (3-) 5-veined, secondary veins parallel... *S. spissa*, p. 158.

1. *Smilax angustiflora* A. DC., Monogr. phan. 1: 67. 1878. Figure 12.1A.

Description. Vine; stems terete, pubescent, sometimes short-tomentose, prickles absent. Leaves simple, to 13 cm long and 5.5 cm wide, ovate-oblong, scattered pilose above, short-pilose below, 7-veined; tip acute; base cordate; margins entire; petioles to 2 cm long, tomentose, bearing two stipular, simple tendrils. Peduncles of staminate umbels tomentose, to 2.5 cm long, somewhat longer than subtending petioles, solitary, axillary. Staminate flowers with pedicels tomentose, to 6 mm long; perianth segments linear, to 9 mm long and 1 mm wide, tomentose but lacking tuft of hair at tip; filaments to 6 mm long; anthers to 2 mm long.
Phenology. Flowers in June. Fruiting times unknown.
Distribution. Endemic to Costa Rica. Elevations unknown.

2. *Smilax candelariae* A. DC., Monogr. phan. 1: 70. 1878. Figure 12.1B.

Description. Vine; stems terete, glabrate to short-pilose; prickles absent. Leaves simple, to 16 cm long and 9 cm wide, unlobed, ovate or oblong, glabrous above, short pilose below, 7-veined; tip acuminate; base cordate, subcordate or sometimes rounded; margins entire; petioles pubescent (at least when young), to 1.5 cm long, bearing two stipular, simple tendrils. Peduncles of fruiting umbels solitary, to 5 mm long, axillary; receptacle globose; bractlets connate forming a cup at pedicel base, pubescent; pedicels hairy, to 8 mm long, swollen at base. Fruit red, to 1.2 cm long and 1 cm wide.
Phenology. Flowers from April through May. Fruits in January and May through June.
Distribution. Endemic to Costa Rica. Found to at least 1,600 m.

3. *Smilax chiriquensis* C.V. Morton, in Woodson & Schery, Ann. Missouri Bot. Gard. 29: 326. 1942. Figure 12.1C.

Description. Vine; stems sharply tetragonal, glabrous, frequently with scattered prickles. Leaves simple, to 19 cm long and 12 cm wide, ovate, glabrous above and below, 7–9-veined; outer veins marginal; secondary veins distinctly reticulate and raised above and below; tip acuminate to apiculate; base cordate to broadly truncate; margins entire; petioles glabrous, to 6 cm long, bearing two stipular, simple tendrils. Peduncles of staminate

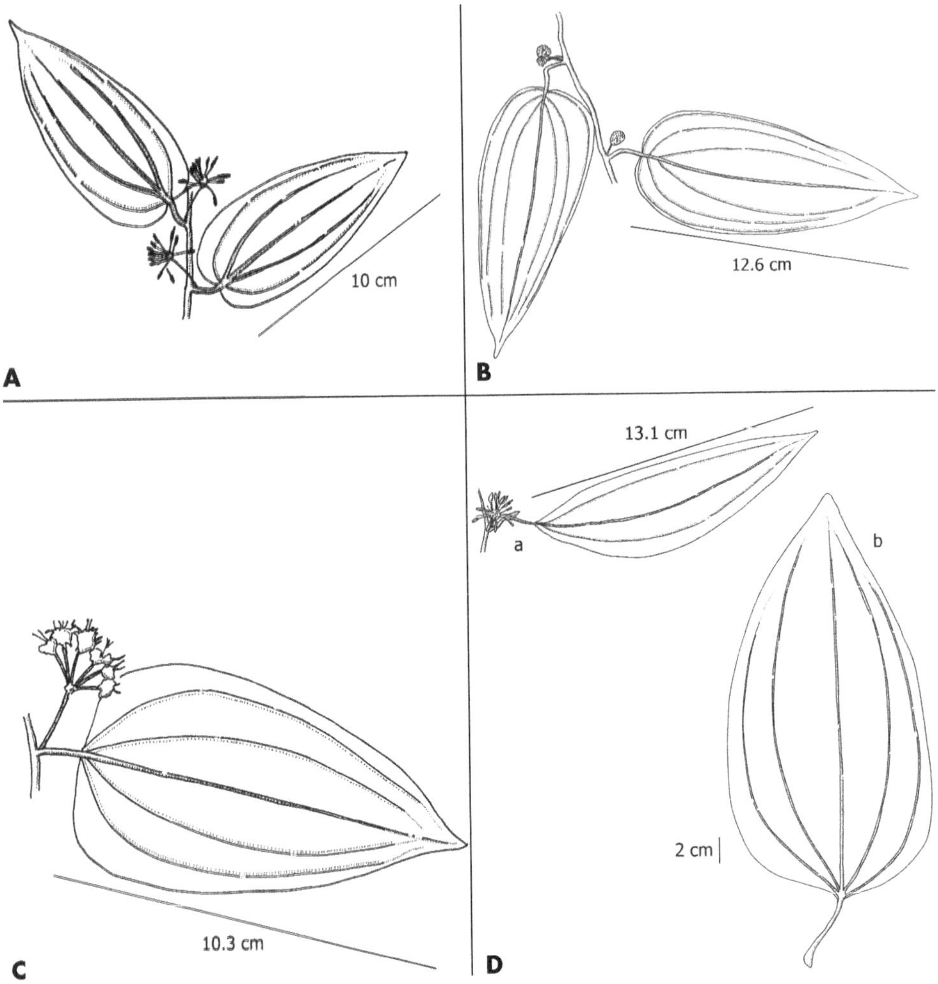

Fig. 12.1. A. *Smilax angustiflora:* leaves and inflorescences (Based on Hoffmann 575, F [photo]). **B.** *Smilax candelariae:* leaves and receptacles (Based on Shank and Molina 4431, US). **C.** *Smilax chiriquensis:* leaf and inflorescence (Based on White 59, US). **D.** *Smilax domingensis:* **(a)** leaf and inflorescence, **(b)** leaf variation (Leaf and inflorescence based on Sperry 943, US; Leaf variation based on Utley and Utley 4810, US).

umbels glabrous, longer than subtending petiole, flattened. Staminate flowers with pedicels glabrous, to 1.1 cm long; perianth segments linear, to 9 mm long and 2 mm wide, glabrous; filaments to 6 mm long; anthers to 2 mm long. Peduncles of fruiting umbels solitary, to 8 mm long, axillary; receptacle globose; pedicels to 1.3 cm long. Fruit bright red, to 1.2 cm long and 0.9 cm wide.

Phenology. Flowers from November through January and in March. Fruits from December through February.

Distribution. Costa Rica and Panama. In Costa Rica found to 1,900 m.

4. ***Smilax domingensis*** Willd., Sp. pl. 4: 783. 1806. Figure 12.1D.

Description. Vine; stems terete, glabrous, frequently with scattered prickles. Leaves simple, to 9 cm long and 5 cm wide, ovate to ovate-lanceolate, glabrous above and below, 5-veined; outer veins marginal; secondary veins distinctly reticulate, raised above and below; tip acuminate; base acute; margins entire; petioles glabrous or with minute, scattered hairs, to 1.6 cm long, bearing two stipular, simple tendrils. Peduncles of staminate umbels glabrous, to 5 mm long, shorter than subtending petiole, flattened; receptacle to 4 mm wide; bractlets glandular-fimbriate, forming a 1 mm-long cup at base of each pedicel. Staminate flowers with pedicels to 7 mm long; perianth segments ligulate, to 7 mm long and 2 mm wide, glabrous, recurved; filaments to 4 mm; anthers to 2 mm. Peduncles of pistillate umbels subterete, to 7 mm long, shorter than subtending petiole; receptacle to 4 mm wide. Pistillate flowers with pedicels to 1 cm long, frequently longer than peduncle; perianth segments oblong-lanceolate, to 4 mm long; staminodia 3, to 1 mm long; styles 3, nearly 2 mm long. Peduncles of fruiting umbels solitary, to 5 mm long, axillary; receptacle globose; pedicels glabrous, to 1.2 cm long, swollen at base. Fruit brownish to dull red, to 1 cm wide, globose.

Phenology. Flowers from April through November. Fruits from June through February.

Distribution. Mexico to Panama. In Costa Rica found from sea level to 2,200+ m.

5. ***Smilax engleriana*** F.W. Apt, Repert. Sp. Nov. Fedde 18: 407. 1922. Figure 12.2A.

Description. Vine; stems terete, glabrous, frequently with scattered prickles. Leaves simple, to 15 cm long and 5 cm wide, lanceolate, glabrous above and below, 5-veined; outer two veins marginal; inner three veins (including midvein) conspicuously impressed above, raised below; secondary veins obscure above and below; tip acuminate; base acute; margins entire; petioles glabrous, to 2.3 cm long, bearing two stipular, simple tendrils. Peduncles of staminate umbels solitary, shorter than subtending petiole, axillary, glabrous; receptacle subglobose, to 4 mm wide. Staminate flowers with pedicels glabrous, to 7 mm long; perianth segments ligulate, to 6 mm long, glabrous or with minute appressed hairs, apical tuft of long hairs absent; filaments nearly 3 mm long; anthers to 2 mm long. Peduncles of pistillate umbels solitary, to 6 mm long, shorter than subtending petiole, axillary, flattened. Pistillate flowers with pedicels to 8 mm long; perianth segments lanceolate, to 4 mm long, recurved; staminodia 6, to 1 mm long, subulate; ovary rugose; styles 3, to 1.2 mm long. Peduncles of fruiting umbels solitary, to 1.5 cm long, axillary, glabrous, disc-shaped at apex, swollen at base. Fruit red to dull purple, to 1 cm wide, globose.

Phenology. Flowers in February and March. Fruits in January.

Distribution. Nicaragua and Costa Rica. In Costa Rica found from 600 to 1,800 m.

6. ***Smilax kunthii*** Killip & C.V. Morton, Publ. Carnegie Inst. Wash. 461: 269. 1936. Figure 12.2B.

Description. Vine; stems terete, glabrous, frequently with scattered prickles. Leaves simple, to 12 cm long and 8 cm wide, ovate, glabrous above and below, 7-veined; outer veins marginal or submarginal; all veins (including the distinctly reticulate secondary veins) noticeably raised above and below; tip acuminate; base rounded or subcordate; margins entire; petioles glabrous, to 3 cm long, bearing two stipular, simple tendrils. Staminate flowers in umbels, solitary or in short racemes, to 1.3 cm long, shorter than subtending petiole, ax-

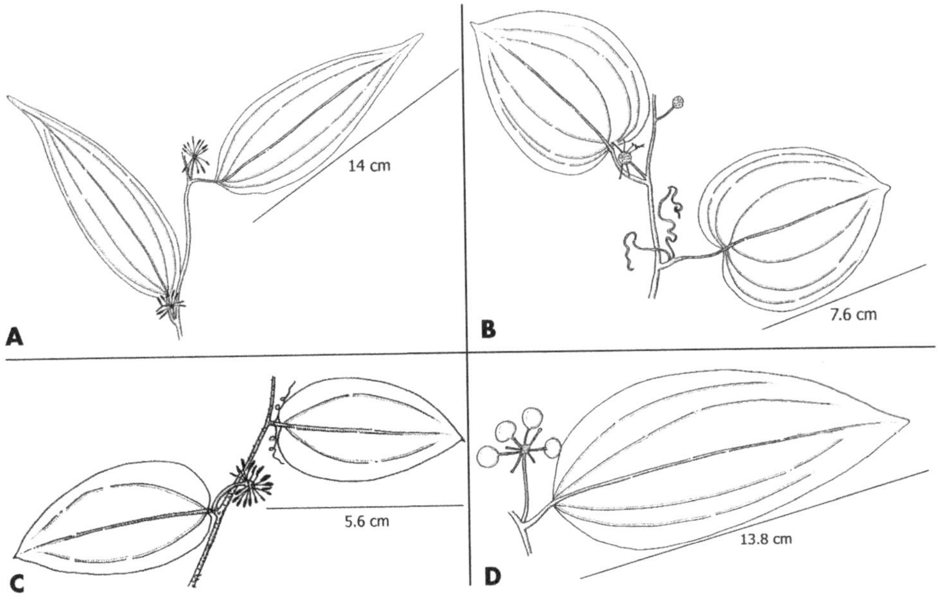

Fig. 12.2. A. *Smilax engleriana:* leaves and inflorescences (Based on Tonduz 11732, US). **B.** *Smilax kunthii:* leaves, receptacles, and tendrils (Based on Standley and Valerio 50770, US). **C.** *Smilax mollis:* leaves, tendrils, and inflorescence (Based on Purpus 2011, US). **D.** *Smilax panamensis:* leaf and infructescence (Based on Pittier 10623, US).

illary, glabrous, subterete; receptacle globose, to 4 mm wide. Staminate flowers with pedicels glabrous, to 7 mm long, shorter or longer than the peduncles; perianth segments lanceolate, to 5 mm long, recurved, glabrous; filaments nearly 3 mm long, anthers to nearly 2 mm long. Peduncles of pistillate umbels glabrous, at least 6 mm long, shorter than subtending petiole, solitary, axillary, flattened. Pistillate flowers with pedicels at least 6 mm long; perianth segments lanceolate, to 4 mm long; staminodia 6, to 2 mm long, subulate; ovary to 2 mm long; styles 3. Peduncles of fruiting umbels solitary, to 2 cm long, shorter than subtending petioles, axillary, glabrous or with a few scattered hairs, terete; receptacle globose, to 4 mm wide; pedicels somewhat swollen at base, to 9 mm long. Fruit green (immature?), at least 5 mm wide, globose.

Phenology. Flowers from March through June and in August and December. Fruits from February through July and in November.

Distribution. Costa Rica to at least Ecuador. In Costa Rica found from near sea level to 2,700 m.

7. *Smilax mollis* Humb. & Bonpl. ex Willd., Sp. pl. 4: 785. 1806. Figure 12.2C.

Description. Vine; stems terete, short pilose to short tomentose; prickles absent. Leaves simple, to 17 cm long and 10 cm wide, ovate to ovate-oblong, somewhat pubescent above when young, soon glabrate, persistently hirsutulous below (at least on major veins), 7-veined; outer veins marginal or submarginal; secondary veins reticulate, raised above and below; tip acute; base cordate; margins entire; petioles densely pubescent, to 1.8 cm long, bearing two stipular, simple tendrils. Staminate flowers in umbels, frequently solitary, to 4 cm long, longer than subtending petiole, axillary, densely hirsute or tomentose, terete; receptacle globose, to 4 mm wide. Staminate flowers with pedicels short-hirsute or infrequently glabrous, to 5 mm long, shorter than the peduncles; perianth segments oblong or oblong-linear, to 5 mm long, recurved, somewhat hairy to glabrous but frequently with an apical tuft of hair; filaments nearly 4 mm long; anthers 1+ mm long. Peduncles of pistillate umbels short-hirsute, to 1.9 cm long, longer than subtending petiole, solitary, axillary, terete, or somewhat flattened; receptacular bractlets short. Pistillate flowers with pedicels to 5 mm long, short-hirsute; perianth segments oblong or linear-oblong, to 3 mm long, recurved, glabrous except for apical tuft of hair; staminodia 3, to 1 mm long. Peduncles of fruiting umbels solitary, to 2.7 cm long, often significantly longer than subtending petioles, axillary, often short-hirsute, terete; receptacle globose, to 4 mm wide; pedicels somewhat swollen at base, to 7 mm long. Fruit yellow or red, to 8 mm wide, globose.

Phenology. Flowers from April through July. Fruits from May through November and in January and February.

Distribution. Mexico to Panama. In Costa Rica found to 2,000 m.

8. *Smilax panamensis* Morong, Bull. Torrey Bot. Club 21: 441. 1894. Figure 12.2D.

Description. Vine; stems terete, glabrous; prickles to 2 cm long. Leaves simple, to 18 cm long and 10 cm wide, ovate, lanceolate, or oblong, glabrous above and below, 7-veined; outer veins marginal or submarginal; secondary veins reticulate; principal veins somewhat impressed above, raised below; secondary veins raised below; tip acuminate; base obtuse; margins entire; petioles glabrous or with a few scattered hairs, to 3 cm long, bearing two stipular, simple tendrils. Peduncles of staminate umbels paired to clustered, longer than subtending petiole, axillary, glabrous, flattened; receptacle nearly 3 mm wide. Staminate flowers with pedicels terete, to 8 mm long, shorter or longer than the peduncles, glabrous; perianth segments ovate-lanceolate, to 6 mm long, straight; anthers longer than filaments, to nearly 3 mm long. Peduncles of pistillate umbels glabrous, to 1.5 cm long, often shorter than subtending petiole, solitary or clustered, axillary, flattened. Pistillate flowers with pedicels to 1.2 cm long, glabrate; perianth segments more or less linear-oblong, to 5 mm long, glabrous; staminodia 6, to 1 mm long; styles 3, to 1.5 mm long. Peduncles of fruiting umbels solitary or clustered, to 2 cm long, often shorter than subtending petioles, axillary, glabrous, flattened; receptacle to 5 mm wide; pedicels swollen at base, to 1.7 cm long. Fruit greenish-white (immature?), to 1.3 cm wide, subglobose.

Phenology. Flowers in April. Fruits in February, from April through June, and August through September.

Distribution. Guatemala to Panama. In Costa Rica found from sea level to 2,700 m.

9. *Smilax spinosa* Mill., Gard. dict. ed. 8, No. 8. 1768. Figure 12.3A.

Description. Vine; stems subterete or angulate, glabrous, frequently with prickles. Leaves simple, to 15 cm long and 8.5 cm wide, ovate to ovate-elliptic, glabrous above and below, 5-veined; principal and secondary veins raised above and below, sometimes with prickles below; secondary veins reticulate; tip acute to rounded; base subcordate; margins entire; petioles glabrous, to 1.7 cm long, bearing two stipular, simple tendrils. Peduncles of staminate umbels solitary, to 8 mm long, shorter than subtending petiole, axillary, flattened; receptacle globose to ellipsoid, to 3 mm wide. Staminate flowers with pedicels glabrous, to 1.3 cm long, longer than the peduncle; perianth segments linear-oblong, to 3 mm long, glabrous or with scattered hairs, lacking apical tuft of hair; filaments to 1 mm long; anthers less than 1 mm long. Peduncles of pistillate umbels glabrous, to 9 mm long, shorter than subtending petiole, solitary, axillary, flattened. Pistillate flowers with pedicels to 9 mm long, glabrate; perianth segments oblong or linear-oblong, to nearly 3 mm long, recurved; staminodia 3 or 6; styles 3. Peduncles of fruiting umbels solitary, to 1.2 cm long, shorter than subtending petioles, axillary, glabrate, terete; receptacle globose, to 4 mm wide; pedicels somewhat swollen at base, to 9 mm long. Fruit black, to 1.2 cm wide, globose.

Phenology. Flowers from March through April. Fruits from December through March and in June and July.

Distribution. Mexico to Panama. In Costa Rica found from sea level to 2,000 m.

10. *Smilax spissa* Killip & C.V. Morton, Publ. Carnegie Inst. Wash. 461: 273. 1936. Figure 12.3B.

Description. Vine; stems terete, glabrous, frequently with prickles. Leaves simple, to 16 cm long and 6 cm wide, oblong, glabrous above and below, 5-veined; principal veins impressed above, raised below; secondary veins more or less parallel, not reticulate, raised below; tip acuminate; base broadly attenuate to rounded; margins entire; petioles glabrous, to 2 cm long, bearing two stipular, simple tendrils. Peduncles of staminate umbels solitary, to 4.5 cm long, longer than subtending petiole, axillary, subterete; receptacle subglobose, to 5 mm wide. Staminate flowers with pedicels obsolete; perianth segments oblong-ovate, to 3 mm long, glabrous to somewhat puberulent; filaments to 1.5 mm long; anthers to 2 mm long. Peduncles of pistillate umbels glabrous, to 1.5 cm long, longer than subtending petiole, solitary, axillary, flattened. Pistillate flowers with pedicels to 5 mm long, glabrous; perianth segments oblong to ovate, to nearly 3 mm long, glabrous, lacking apical tuft of hair. Peduncles of fruiting umbels solitary, to 2.3 cm long, longer than subtending petioles, axillary, terete, glabrous; receptacle globose, to 4 mm wide; pedicels somewhat swollen at base, to 7 mm long. Fruit red, to 1.5 cm wide, globose.

Phenology. Flowers in February and March. Fruits from May through July, in September, and from November through February.

Distribution. Costa Rica and Panama. In Costa Rica found from sea level to 1,100+ m.

11. *Smilax subpubescens* A. DC., Monogr. phan. 1: 69. 1878. Figure 12.3C.

Description. Vine; stems somewhat tetragonal, tomentose when young, soon glabrous, prickles absent. Leaves simple, to 24 cm long and 19 cm wide, ovate, reddish tomentose to glabrous above and below, 7–11-veined; secondary veins reticulate; tip acuminate; base cordate to rounded or truncate; margins entire; petioles tomentose to glabrous, 1.7–7 cm long, bearing two stipular, simple tendrils. Peduncles of staminate umbels solitary, to 3.5 cm long, shorter than subtending petiole (rarely longer), axillary, flattened, reddish-tomentose; receptacle globose, to 5 mm wide; bractlets many, pubescent. Staminate flowers with pedicels reddish or brownish short-tomentose, to 1.5 cm long, shorter than the peduncle; perianth segments oblong to oblong-linear, to 6 mm long, sparsely to densely pubescent; filaments nearly 4 mm long; anthers to 2 mm long. Peduncles of pistillate umbels short-tomentose or pilose, to 1 cm long, shorter than subtending petiole, solitary, axillary, flattened; receptacular bractlets short. Pistillate flowers with pedicels to 6 mm long; perianth segments lanceolate, to nearly 4 mm long, short-tomentose; staminodia 6. Peduncles of fruiting umbels frequently solitary, to 2.5 cm long, shorter than subtending petioles, axillary, short-tomentose to pilose, terete; receptacle globose, to 4 mm wide; pedicels to 1.8 cm long. Fruit orange to reddish, to 8 mm wide, subglobose.

Phenology. Flowers from March through June. Fruits from January through May.

Distribution. Mexico to Panama. In Costa Rica found to 2,900 m.

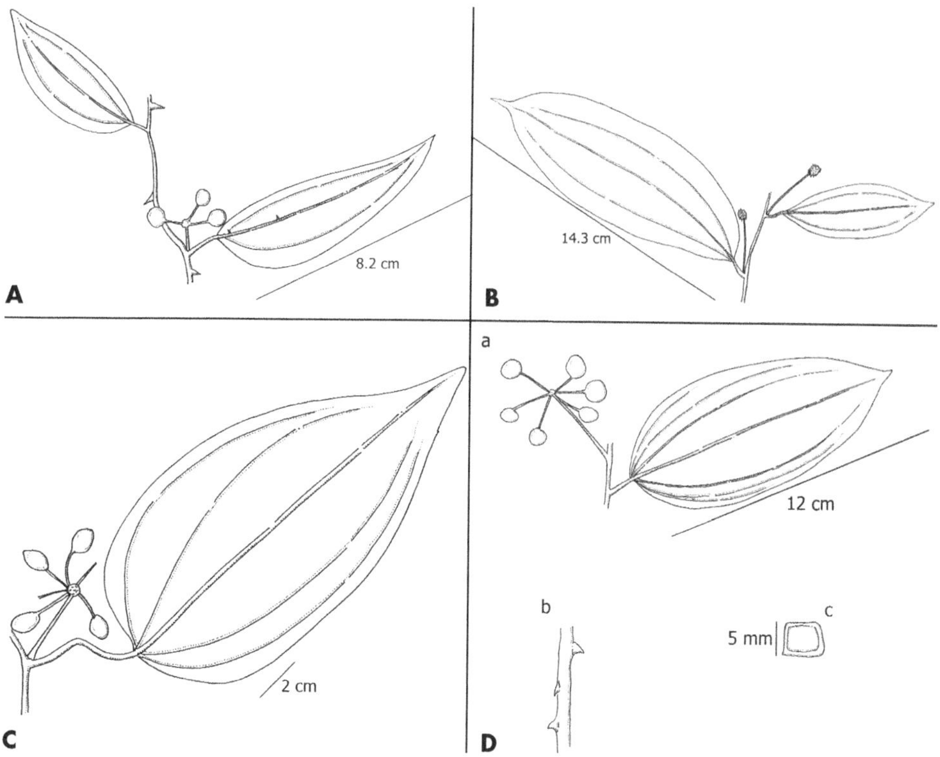

Fig. 12.3. A. *Smilax spinosa:* leaves and infructescence (Based on Calderón 1128, US). **B.** *Smilax spissa:* leaves and receptacles (Based on Pittier 407, US). **C.** *Smilax subpubescens:* leaf and infructescence (Based on Tonduz 11803, US). **D.** *Smilax vanilliodora:* **(a)** leaf and infructescence, **(b)** stem with prickles, **(c)** cross-section of stem (Based on Standley and Valerio 49003, US).

12. *Smilax vanilliodora* F.W. Apt, Repert. Sp. Nov. Fedde 18: 416. 1922. Figure 12.3D.

Synonymy. *Smilax standleyi* Killip & Morton, Publ. Carnegie Inst. Wash. 461: 280. 1936.

Description. Vine; stems tetragonal, glabrous, frequently with prickles. Leaves oblong or oblong-lanceolate, to 21 cm long and 10 cm wide, glabrous above and below, 7–9-veined; principal veins raised above and below; secondary veins reticulate, raised above and below; tip acuminate; base cuneate to subcordate; margins entire; petioles glabrous, to 3.5+ cm long, bearing two stipular, simple tendrils. Peduncles of staminate umbels solitary, to 7 cm long, longer than subtending petiole, axillary; receptacle to 3 mm wide. Staminate flowers with pedicels glabrous, to 1.3 cm long, shorter than the peduncle; perianth segments lanceolate to ovate, to 5 mm long, glabrous to somewhat puberulent; anthers longer than the filaments. Peduncles of pistillate umbels glabrous to somewhat puberulent, to 4 cm long, longer than subtending petiole, solitary or in axillary branchlets; receptacle globose, to 4 mm wide. Pistillate flowers with pedicels glabrous, to 2 cm long; perianth segments lanceolate, to 4 mm long, glabrous, lacking apical tuft of hair. Peduncles of infructescent umbels solitary, to 4.5 cm long, longer than subtending petioles, axillary, glabrous; receptacle globose, to 4 mm wide; pedicels to 2 cm long. Fruit red, to 1.2 cm wide, globose to trilobate.

Phenology. Flowers from October through March and in May. Fruits from December through February and April through September.

Distribution. Endemic to Costa Rica. Found from sea level to 1,900 m.

Notes. When present, prickles found below on the leaf midrib often curve towards the leaf base. Prickles of the closely related *S. regelii* often curve toward the leaf tip.

13. *Smilax velutina* Killip & C.V. Morton, Publ. Carnegie Inst. Wash. 461: 283. 1936.

Figure 12.4.

Description. Vine; stems terete, yellowish velutinous when young, sometimes glabrate with age, prickles absent. Leaves simple, to 15 cm long and 9 cm wide, ovate to ovate-oblong, tomentulose above when young but soon glabrate, yellowish velutinous to subtomentose below (at least on main veins); 7-veined; principal veins generally impressed above, conspicuously raised below; secondary veins reticulate; tip acuminate; base cordate to subcordate; margins entire; petioles densely tomentose or velutinous, to 2.5 cm long, bearing two stipular, simple tendrils. Peduncles of staminate umbels solitary, to 4+ cm long, shorter or longer than subtending petiole, axillary, terete, velutinous; receptacle globose, to 5 mm wide; bractlets pubescent, many. Staminate flowers with pedicels villous, to 4 mm long, shorter than the peduncle; perianth segments linear, to 6 mm long, pubescent externally; filaments nearly 4 mm long; anthers slightly more than 1 mm long. Peduncles of pistillate umbels pubescent, to 2.5 cm long, frequently slightly longer than subtending petiole, solitary, axillary. Pistillate flowers with pedicels pubescent, to 1.2 cm long; perianth segments lanceolate to oblong, to nearly 5 mm long, glabrous internally, short-pilose externally, often reflexed; styles 3. Peduncles of fruiting umbels frequently solitary, to 1.5 cm long, shorter or longer than subtending petiole, axillary, velutinous, terete; receptacle velutinous, to 4 mm wide; pedicels pubescent, to 8 mm long. Fruit greenish-yellow to orange (to black?), to 6 mm wide, subglobose.

Phenology. Flowering times unknown. Fruits in July.

Distribution. Mexico to Costa Rica. Elevation range in Costa Rica unknown.

Notes. No specimens from our area were seen at either CR, F, MO, or US. The species has been included here based on Haber (1991).

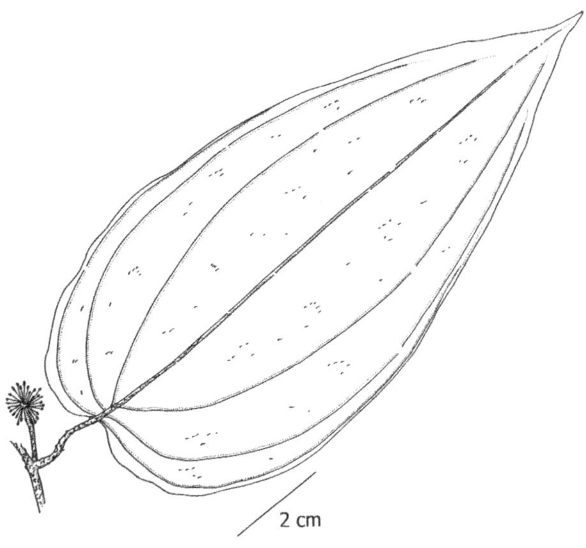

Fig. 12.4. *Smilax velutina:* leaf and inflorescence (Based on Breedlove 11944, US).

13
VITACEAE

Comprising roughly 12 genera and 700 species (Chant 1978), the Vitaceae is distributed throughout tropical, subtropical, and temperate zones. Two genera are found in our area, *Cissus* and *Vitis*. Ten of the 12 species in this family in Costa Rica occur in our area.

The family is readily identified vegetatively by the tendrils that develop from terminal buds and appear opposite the leaves at the nodes. Leaves are alternate, simple or trifoliolate (in our area) and often serrate or crenate. *Cissus* has both trifoliolate and simple leaves, frequently with smooth grayish and blotchy bark. In flower, *Cissus* and *Vitis* can be distinguished by the structure of the inflorescence (a cyme in *Cissus* but a panicle in *Vitis*). In addition, flowers of *Cissus* are 4-merous whereas flowers of *Vitis* are 5-merous.

The only native species of *Vitis* in our area, *Vitis tiliifolia*, is identified by densely floccose-tomentose leaves and shredding, dark bark. *Ampelocissus costaricensis*, found in lowlands, resembles *V. tiliifolia* in overall leaf shape, but can be distinguished by the sparse floccose leaf trichomes that abcise on older leaves.

KEY TO VITACEAE

1. Leaves compound... *Cissus*, p. 161.
1. Leaves simple
    2. Leaves not floccose-tomentose below; inflorescence in cymes; flowers 4-merous... *Cissus*, p. 161.
    2. Leaves densely floccose-tomentose below; inflorescence in panicles; flowers 5-merous... *Vitis*, p. 168.

1. GENUS *CISSUS*

Although the genus *Cissus* is easily identified vegetatively by the leaves that are either compound or not densely floccose-tomentose below, specific identification is more difficult in some cases.

KEY TO *CISSUS*

1. Leaves simple
    2. Leaves strongly dimorphic... *C. biformifolia*, p. 162.
    2. Leaves not strongly dimorphic
        3. Leaves narrowly elliptic to lanceolate; bases rounded to acute... *C. brevipes*, p. 164.
        3. Leaves broadly ovate; bases truncate to cordate, sometimes rounded
            4. Pedicels hirtellous... *C. cacuminis*, p. 164.
            4. Pedicels glabrous... *C. verticillata*, p. 164.
1. Leaves trifoliolate
    5. Terminal leaflets more or less rhombic
        6. Fruits ca. 1 cm long and wide; petals greenish-yellow (immature?) or red, strigillose externally... *C. rhombifolia*, p. 165.

6. Fruits 2.4–3.5 cm long, 1.7–2.3 cm wide; petals greenish-white, papillose externally... *C. ulmifolia*, p. 167.
5. Terminal leaflets ovate, obovate, elliptic, or lanceolate
   7. Stems tetragonal
      8. Lateral leaflets sessile or nearly so; leaves crenate, or if somewhat serrate the teeth strongly incurved; peduncles 4–10 cm long... *C. erosa*, p. 164.
      8. Lateral leaflets with petiolules 3–13 mm long; leaves serrate with teeth spreading, or at least not incurved; peduncles less than 2.7 cm long... *C. microcarpa*, p. 165.
   7. Stems terete
      9. Terminal leaflet 4–10 cm long; petiolule 0.5–2 cm long; margins serrate; flowers red to reddish-orange... *C. microcarpa*, p. xx.
      9. Terminal leaflet 2–4 cm long, sessile; margins crenate; flowers greenish... *C. trianae*, p. 165.

1. *Cissus biformifolia* Standl., Publ. Field Mus. Nat. Hist., Bot. Ser. 4: 225. 1929. Figure 13.1A.

Synonymy. *Cissus cardiophylla* Standl., Publ. Field Mus. Nat. Hist., Bot. Ser. 4: 226. 1929; *Cissus pseudosicyoides* Croat, Ann. Missouri Bot. Gard. 60: 564. 1973, synon. nov.

Description. Vine or liana; stems angulate to terete, pilose when young, soon glabrous. Leaves simple, dimorphic; first form narrowly ovate, elliptic, or obovate, glabrous, 4–10 cm long, 2–5 cm wide, acuminate at tip, cuneate at base; second form more broadly ovate, 5–14 cm long, 4–12 cm wide, scattered minute T-shaped trichomes below; tip acuminate; base cordate; margins serrate near tips, entire at base; petioles 3–8 cm long. Tendrils bifid to many-branched, borne opposite leaves. Inflorescence cymose, to 5 cm long, borne opposite leaves; peduncles 2–3 cm long, appressed pilose when young. Flowers red to dark red; pedicel 1–2 mm long, scattered appressed pilose; calyx 4-parted, cupular; petals 4, 3–4 mm long, free, ovate to oblong. Fruits purplish-black, 1–1.3 cm long, 0.8–1 cm wide, obovoid.

Phenology. Flowers and fruits throughout the year.

Distribution. Belize through Peru. In Costa Rica found from sea level to 1,400 m.

Notes. Dodson and Gentry (1978) and Croat (1973) separated *C. biformifolia* and *C. pseudosicyoides* by the following subkey characters:

Fig. 13.1. A. *Cissus biformifolia:* **(a)** leaves, tendrils, and inflorescence; **(b)** flower bud; **(c)** flower with two stamens removed; **(d)** fruit (Leaves, tendrils, and inflorescence based on Gentry 48543A, MO; Flowers based on Todzia 1159, DUKE; Fruit based on Hammel 11681, MO). **B.** *Cissus brevipes:* **(a)** leaf, **(b)** fruit, **(c)** flower bud, **(d)** flower with petals and stamens removed (Based on Skutch 2256, MO). **C.** *Cissus cacuminis:* **(a)** leaves, tendril, and inflorescence; **(b)** flower bud; **(c)** flower with petals and stamens removed (Leaves, tendril, and inflorescence based on Breedlove 11217, DUKE; Flowers based on

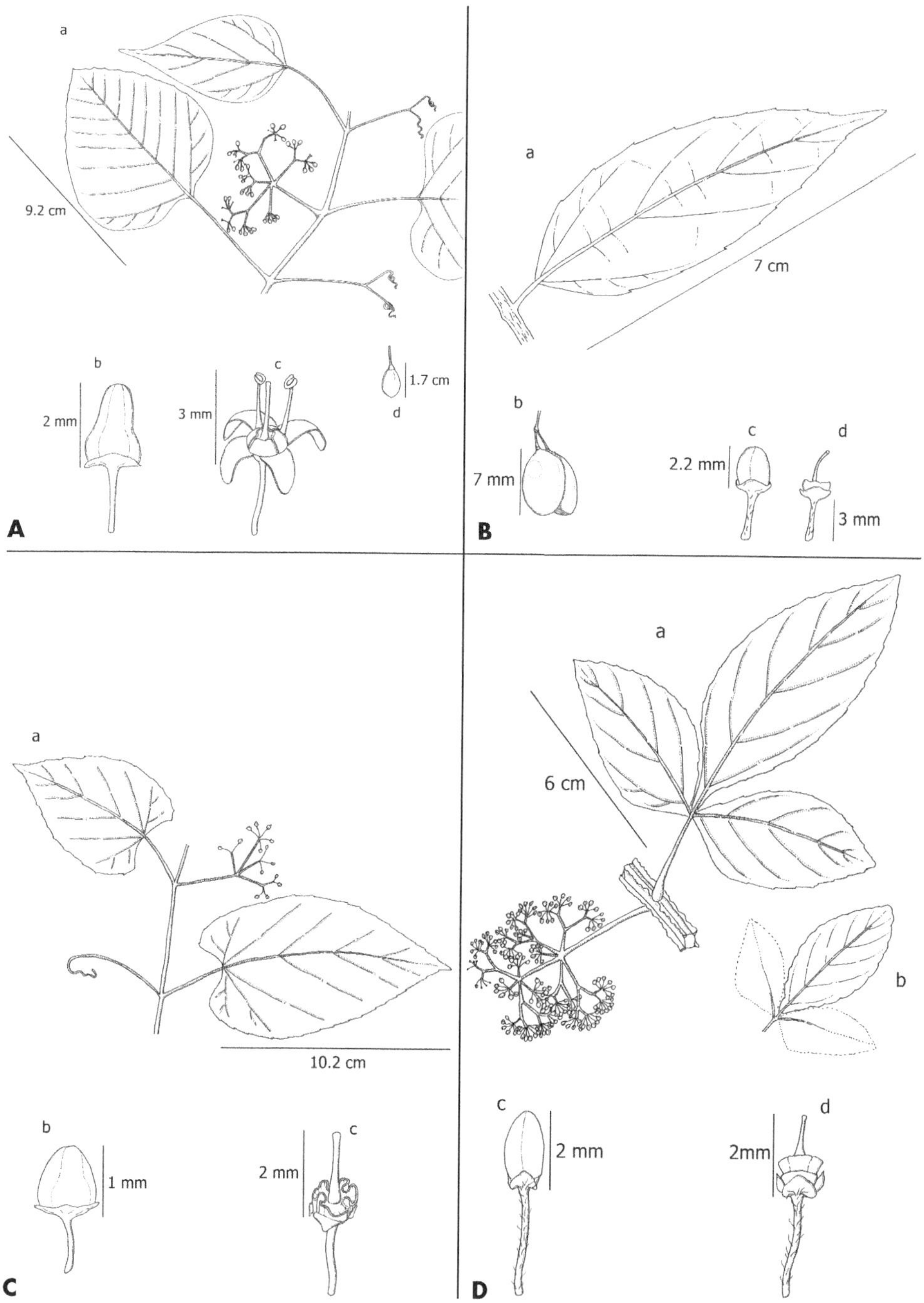

Breedlove 20787, DUKE). **D.** *Cissus erosa:* **(a)** leaf and inflorescence, **(b)** leaf variation, **(c)** flower bud, **(d)** flower with petals and stamens removed (Leaf and inflorescence based on Wilbur 21014, DUKE; Flowers based on Austin 9, DUKE).

Subkey for *C. biformifolia* and *C. pseudosicyoides*

1. Inflorescence appressed-pubescent; pedicels with whitish trichomes; calyx often with sparse, whitish, T-shaped trichomes, wider than the bud, enveloping to half of the base of the corolla; buds 3–4 mm long; fruit 8–10 mm wide... *C. biformifolia*, p. 162.
1. Inflorescence erect-pubescent or glabrous; pedicels with dark trichomes; calyx glabrous, narrower than the bud, shorter than the corolla; buds 1.5–2 mm long; fruits to 6 mm wide... *C. pseudosicyoides*, p. 162.

While these characters seem distinct in a key, none gave consistent results when applied to herbarium specimens. Consequently, *Cissus pseudosicyoides* is herein reduced to synonymy beneath *C. biformifolia.*

2. *Cissus brevipes* C.V. Morton & Standl., Fieldiana, Bot. Ser. 18: 653. 1937. Figure 13.1B.

Description. Vine or liana; stems subterete to terete. Leaves simple, to 12 cm long and 4 cm wide, narrowly elliptic to lanceolate; tip acuminate; base acute to rounded; margins serrate; petioles 1–1.2 cm long. Tendrils simple (?), borne opposite leaves. Inflorescence cymose, borne opposite leaves, puberulent to glabrate. Flowers with pedicels ca. 3 mm long, appressed-puberulous; calyx patelliform, to 0.6 mm long and 1.7 mm wide; petals 4, to 2.5 mm long. Fruits subglobose to obovoid, to 6 mm wide.

Phenology. Flowers in January and March. Fruiting times unknown.

Distribution. Costa Rica to Ecuador. In Costa Rica found from 600 to 950 m.

Notes. Some authors (e.g., Elias 1968) reduced *C. brevipes* to synonymy with *C. verticillata.* But we recognize both species because the type specimen of *C. brevipes* (Skutch 2256, MO) does not resemble any *C. verticillata* we have seen from Costa Rica. Indeed, the leaf of *C. brevipes* more closely resembles *C. biformifolia.* It can be distinguished from the latter by having monomorphic leaves with petioles only to 1.2 cm long (frequently over 3 cm long in *C. biformifolia*).

3. *Cissus cacuminis* Standl., Pub. Field Mus. Nat. Hist., Bot. Ser. 17: 375. Figure 13.1C.

Description. Vine or liana; stems subterete to terete, glabrous to sparsely pilose. Leaves simple, 6–12 cm long, 4–7 cm wide, ovate to broadly ovate, glabrous to scattered pubescent above and below; tip acuminate; base cordate; margins serrate; petioles 1.7–5 cm long. Tendrils simple (?), borne opposite leaves. Inflorescence cymose, to 3.5 cm long, borne opposite leaves; peduncles to 4 cm long. Flowers greenish-yellow; pedicels 2–4 mm long, appressed-pilose; calyx patelliform, to 1 mm wide; petals 4, at least 1 mm long. Fruits subglobose to obovoid, to 6 mm wide.

Phenology. Flowers in November in Chiapas, February and June to August in Honduras, September to June in Nicaragua, and October in Costa Rica. Fruits in January in Chiapas.

Distribution. Mexico to Costa Rica. In Costa Rica found from sea level to 1,740 m.

4. *Cissus erosa* Rich., Actes Soc. Hist. Nat. Paris 1: 106. 1792. Figure 13.1D.

Synonymy. *Cissus salutaris* Kunth in H.B.K., Nov. Gen. sp. pl. 5: 225. 1821; *Vitis salutaris* Baker in Mart., Fl. Bras. 14: 211, pl. 52. 1871.

Description. Vine or liana; stems 4-angular, often winged, scattered pilose to glabrate. Leaves trifoliolate, glabrous to sparsely pubescent above and below; terminal leaflet ovate to elliptic, 6–16 cm long, 3–8 cm wide, acuminate at tip, cuneate at base; lateral leaflets asymmetrical, 3–12 cm long, 3–6 cm wide, acute at tip, subcuneate to rounded at base; margins crenate, or if somewhat serrate the teeth distinctly incurved; petioles 1.8–6.4 cm long, sometimes winged. Tendrils simple (?), borne opposite leaves. Inflorescence cymose-corymbiform, borne opposite leaves; peduncles 4.5–11 cm long. Flowers red, 2–5 mm long; pedicels villous; calyx cupular, with short or obscure lobes. Fruits orbicular, to 7 mm wide, purplish-black.

Phenology. Flowers and fruits throughout the year.

Distribution. Mexico to Venezuela, the Guianas, Colombia to Bolivia; West Indies. In Costa Rica found from sea level to 750+ m.

Notes. Called *mano de sapo* in Ecuador (Dodson et al. 1985) and *caro de tres hojas* in Puerto Rico (Acevedo-Rodriguez and Woodbury 1985).

5. *Cissus microcarpa* Vahl, Eclog. amer. 1: 16. 1796. Figure 13.2A.

Synonymy. *Vitis microcarpa* (Vahl) Donn. Sm., Enum. pl. guatem. 3: 19. 1893.

Description. Vine or liana; stems subterete to angular, lacking wings. Leaves trifoliolate, sparsely pubescent to velutinous below on veins; terminal leaflet elliptic to ovate-elliptic, 4–10 cm long, 2–6.5 cm wide, acuminate at tip, cuneate at base; lateral leaflets asymmetrical, to 8 cm long and 5 cm wide; margins serrate to mucronate; petioles 2–5.5 cm long. Tendrils simple to many-branched, disk-tipped, borne opposite leaves. Inflorescence cymose-corymbiform, to 8 cm long, borne opposite leaves; peduncles to 3 cm long, pubescent. Flowers red; calyx minute, cupular; corolla 4-parted, lobes ovate. Fruits obovoid, to 1 cm long and 0.6 cm wide.

Phenology. Flowers and fruits in December, and from March to September.

Distribution. Mexico to Bolivia; Greater Antilles. In Costa Rica found from sea level to 2,400 m.

6. *Cissus rhombifolia* Vahl, Eclog. amer. 1: 11. 1796. Figure 13.2B.

Description. Vine or liana; stems subangular to terete, winged or not, densely pubescent with glandular and eglandular hairs. Leaves trifoliolate, densely pubescent to villous; terminal leaflet rhombic, 8–12 cm long, 6–8 cm wide, acuminate at tip; lateral leaflets asymmetrically ovate to elliptic, 6–10 cm long, 5–7 cm wide, acuminate at tip, obliquely cuneate at base; petioles 2–8 cm long, winged or not. Tendrils many-branched, disc-tipped, borne opposite leaves. Inflorescence cymose-corymbiform, borne opposite leaves. Flowers yellow, greenish-yellow, or red; pedicels pubescent, 2.5–8 mm long; calyx bowl-shaped, lobes obscure. Fruits obovoid, to 1 cm long, blackish.

Phenology. Flowers and fruits in January and from April through November.

Distribution. Mexico to Bolivia; West Indies. In Costa Rica found from sea level to 2,350 m.

Notes. Croat (1978) described Panamanian specimens with "simple tendrils." However, most Costa Rican specimens we examined have many-branched, disc-tipped tendrils. This species is called *xtabcanil* in Mayan (Standley 1930).

7. *Cissus trianae* Planch., Monogr. phan. 5: 555. 1887. Figure 13.2C.

Synonymy. *Cissus martiniana* Woodson & Seibert, Ann. Missouri Bot. Gard. 24: 191. 1937.

Description. Vine or liana; stems subangular to terete, glabrous to scattered pilose when young, generally glabrous with age. Leaves trifoliolate, glabrous to sparsely pubescent

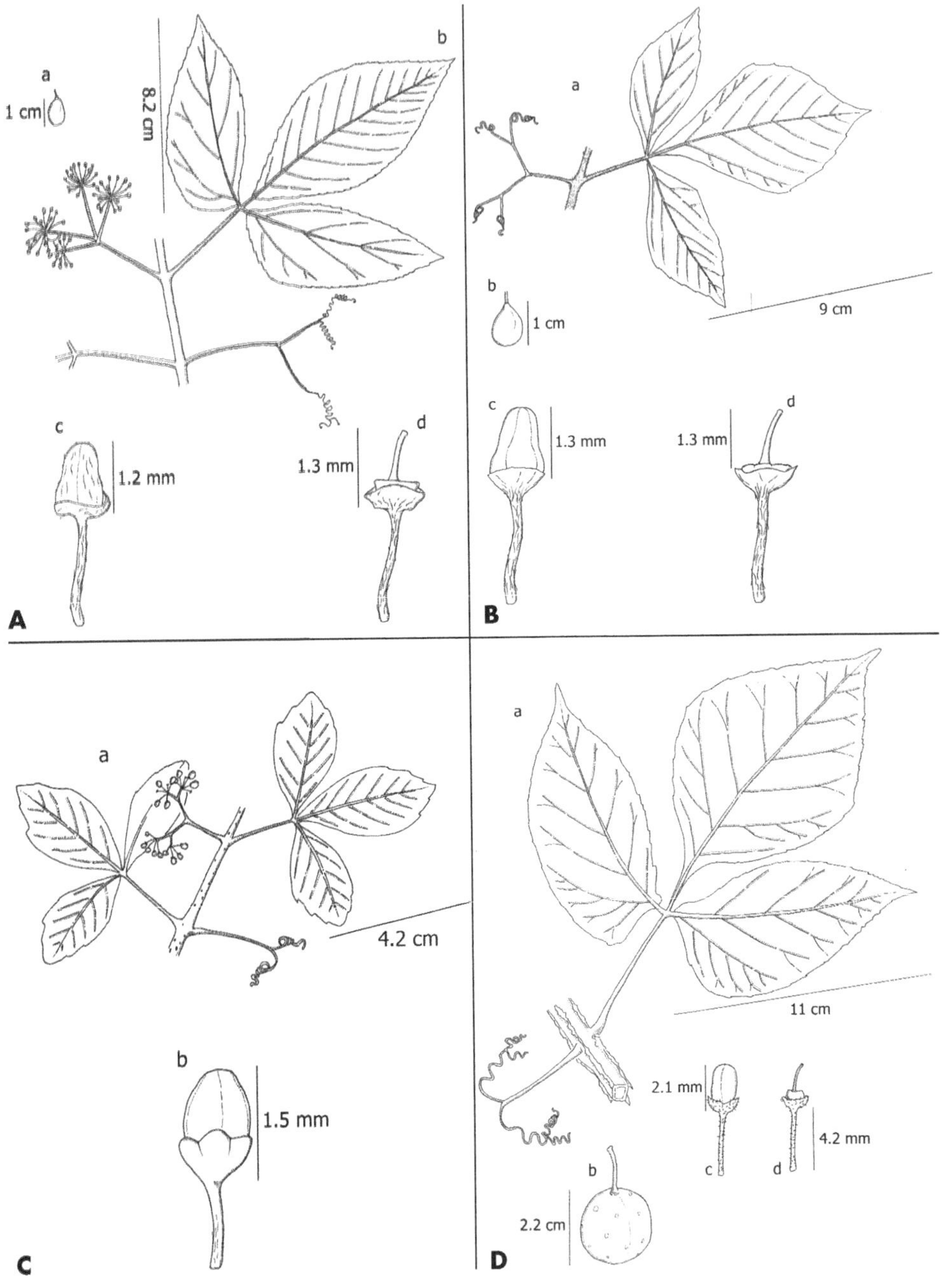

Fig. 13.2. A. *Cissus microcarpa:* **(a)** fruit; **(b)** leaf, tendril, and inflorescence; **(c)** flower bud; **(d)** flower with petals and stamens removed (Leaf, tendril, inflorescence, and fruit based on Wilbur 27227, DUKE; Flowers based on Almeda, Jr. 645, DUKE). **B.** *Cissus rhombifolia:* **(a)** leaf and tendril, **(b)** fruit, **(c)** flower bud, **(d)** flower with petals and stamens removed (Leaf, tendril, and fruit based on Opler 1899, MO; Flowers based on MacDougal 1395, DUKE). **C.** *Cissus trianae:* **(a)** leaves, tendril, and inflorescence; **(b)** flower bud (Leaves, tendril, and inflorescence based on Garwood et al. 1384, MO; Flower bud based on Hatheway 1355, DUKE). **D.** *Cissus ulmifolia:* **(a)** leaf and tendril, **(b)** fruit, **(c)** flower bud, **(d)** flower with petals and stamens removed (Leaf, tendril, and fruit based on Gentry et al. 31148, MO; Flowers based on Grayum and Schatz 5259, MO).

above and below; terminal leaflet sessile, 2–4 cm long, 1.2–2 cm wide, elliptic to obovate, acute to obtuse at apex, cuneate at base, mucronate; lateral leaflets sessile, 1.5–4 cm long, 0.7–2 cm wide, elliptic to obovate, asymmetrically cuneate at base, mucronate; margins denticulate to serrulate; petioles 1–5 cm long. Tendrils bifid to many-branched, disc-tipped (?), borne opposite leaves. Inflorescence cymose-corymbiform, borne opposite leaves; peduncles glabrous or with scattered tomentum, 1–3 cm long. Flowers green; pedicel glabrous, 2–4 mm long; calyx 4-parted; calyx lobes ovate to subreniform; petals 4, to 1.5 mm long, ovate to oblong, acute at apex. Fruits globose, ca. 1 cm wide.

Phenology. Flowers and fruits throughout the year.

Distribution. Guatemala to Panama. In Costa Rica found from 1,100 to 2,470 m.

8. *Cissus ulmifolia* (Baker) Planch., Monogr. phan. 5: 552. 1887. Figure 13.2D.

Synonymy. *Vitis ulmifolia* Baker, Fl. Bras. 14: 213. 1871.

Description. Vine or liana; stems subangular to terete, frequently winged, pubescent with glandular and eglandular hairs. Leaves trifoliolate, pubescent to villous; terminal leaflet rhombic, to 14 cm long and 11 cm wide, acuminate at apex; lateral leaflets asymmetrically ovate to elliptic, 6–10 cm long, 5–7 cm wide, acuminate at apex, obliquely cuneate at base; petioles 2.7–10 cm long, frequently winged. Tendrils many-branched, borne opposite leaves. Inflorescence cymose-corymbiform, borne opposite leaves. Flowers greenish-white; pedicels pubescent, to 4.5 mm long; calyx bowl-shaped; calyx lobes obscure; corolla 4-parted; corolla lobes oblong to ovate, to 2.3 mm long, papillose externally. Fruits ovoid to subglobose, 2.4–3.5 cm long, 1.7–2.3 cm wide.

Phenology. Flowers in May and August. Fruiting times unknown.

Distribution. Costa Rica to Bolivia. In Costa Rica found from sea level to 1,050+ m.

9. *Cissus verticillata* (L.) Nicolson & C.E. Jarvis, Taxon 33: 727. 1984. Figure 13.3A.

Synonymy. *Cissus sicyoides* L., Syst. nat. ed. 10: 897. 1759; *C. elliptica* Schltdl. & Cham., Linnaea 5: 221. 1830; *C. obtusata* Benth., Bot. voy. Sulphur 77. 1844; *Vitis sicyoides* Morales in Poey, Repert. 1: 206. 1866.

Description. Vine or liana; stems terete, villous to glabrate. Leaves simple, 7–12 cm long, 5–8 cm wide, ovate to oblong-ovate, villous (especially on veins) to glabrous below; tip acuminate to acute; base obtuse, truncate, subcordate or cordate; margins serrate; petioles 4–6 cm long. Tendrils bifid, borne opposite leaves. Inflorescence cymose, borne opposite leaves or terminal, umbelliform; peduncles to 6 cm long. Flowers pale yellow to red, to 3 mm long; pedicels glabrous, to 3 mm long; calyx 4-parted, glabrous. Fruits obovoid to subglobose, to 6 mm wide, black.

Phenology. Flowers and fruits throughout the year.

Distribution. Central Florida and Mexico to Argentina, Galapagos, and West Indies. In Costa Rica found from sea level to 1,100+ m.

Notes. Croat (1978) wrote that Panamanian specimens usually have simple tendrils. Acevedo-Rodriguez and Woodbury (1985) described Puerto Rican specimens with both simple and bifid tendrils. The majority of Costa Rican specimens we examined have bifid tendrils.

The species is called possum grape in Florida (Wunderlin 1982), *bejuco de agua* in Ecuador (Dodson et al. 1985), *Sanalotodo* in Guatemala (Standley 1930), *bejuco de caro* in Puerto Rica (Acevedo-Rodriguez and Woodbury 1985), and *bejuco loco* and *uru cimarrona* in Panama (Croat 1978).

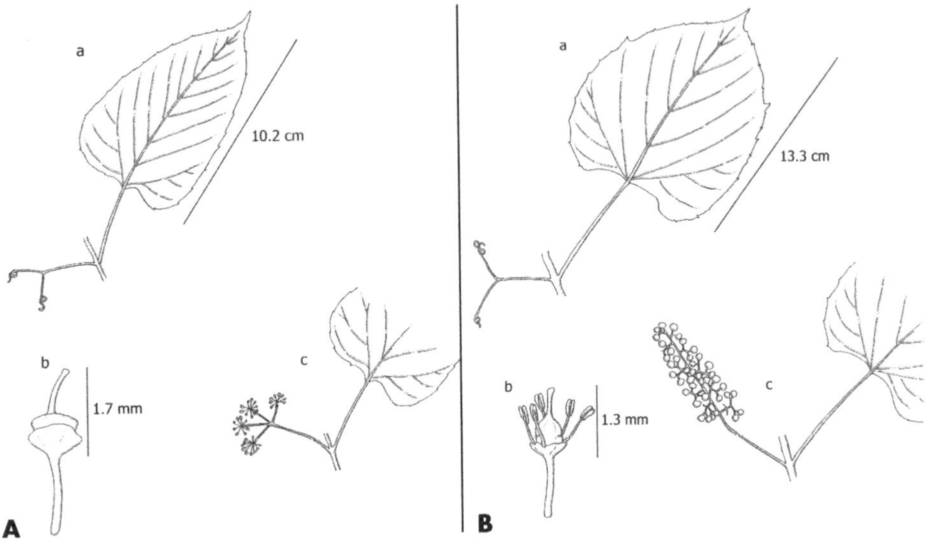

Fig. 13.3. A. *Cissus verticillata:* **(a)** leaf and tendril, **(b)** flower with petals and stamens removed, **(c)** inflorescence (Leaf, tendril, and inflorescence based on Cordero 77, MO; Flower based on Burger et al. 10530, DUKE). **B.** *Vitis tiliifolia:* **(a)** leaf and tendril, **(b)** flower, **(c)** inflorescence (Leaf and inflorescence based on Blanco 23, MO; Flower based on Wilbur 34009, DUKE).

2. GENUS *VITIS*

The genus *Vitis* is represented by only one species in Costa Rica, *V. tiliifolia*. Vegetatively, the species is distinguished by its densely floccose-tomentose leaf undersurfaces. The dark-brown, shredding bark of older lianas is also typical.

1. *Vitis tiliifolia* Humb. & Bonpl. ex Roem. & Schult. in L., Syst. veg. 5: 320. 1819. Figure 13.3B.

Synonymy. *Vitis caribaea* DC., Prodr. 1: 634. 1824.

Description. Liana; stems striate, angulate to terete, densely floccose-tomentose when young, sometimes glabrate with age. Leaves simple, 8–15 cm long, 6–13 cm wide, slightly 3-lobate, ovate-cordate to orbicular, densely floccose-tomentose below; tip acuminate; base cordate or infrequently truncate; margins dentate to serrate; petioles 2–14 cm long. Tendrils forked, borne opposite leaves. Inflorescence paniculate, to 22 cm long, borne opposite leaves; peduncles 5–8 (14) cm long, sparsely to densely floccose-tomentose. Flowers pale yellow to greenish yellow, to 2 mm long; calyx bowl-shaped, margins undulate; petals 5; petal lobes more or less oblong or oblong-ovate; stamen 5. Fruits globose, 4–10 mm wide, purplish-black.

Phenology. Flowers and fruits throughout the year.

Distribution. West Indies, Mexico to Venezuela and Ecuador. In Costa Rica found from sea level to 1,300+ m.

Notes. Sour but edible, the fruits are used more often in Panama for the production of vinegar and wine than for eating fresh (Elias 1968). The stems contain potable water. The species is known throughout its range as *uva* or *bejuco de agua*.

LITERATURE CITED

Preface

Bridson, G. D. R., and E. R. Smith. 1991. *Botanico-Periodicum Huntianum/Supplementum*. Pittsburgh, PA: Hunt Institute for Botanical Documentation.

Brummitt, R. K., and C. E. Powell. 1992. *Authors of Plant Names*. Kew, United Kingdom: Royal Botanic Gardens.

Harris, J. G., and M. W. Harris. 1994. *Plant Identification Terminology, An Illustrated Glossary*. Spring Lake, UT: Spring Lake Publishing.

Holmgren, P., N. H. Holmgren, and L. C. Barnett. 1990. *Index Herbariorum*. New York: New York Botanical Garden.

Krings, A. 1997. *An Illustrated Guide to the Tendrillate Lianas and Vines from the Mountains of Costa Rica*. Unpublished M.S. Thesis, Department of Forestry, College of Forest Resources. Raleigh: North Carolina State University.

Lawrence, G. H. M., A. F. G. Buchheim, G. S. Daniels, and H. Dolezal. 1968. *Botanico-Periodicum-Huntianum*. Pittsburgh, PA: Hunt Botanical Library.

Stafleu, F. A., and R. S. Cowan. 1976. *Taxonomic Literature: A Selective Guide to Botanical Publications and Collections with Dates, Commentaries, and Types*. Utrecht, Netherlands: Bohn, Scheltema, and Holkema.

Chapter 1

Burger, W. C. 1985. Why are there so many kinds of flowering plants in Costa Rica? Pp. 125–136 in *The Botany and Natural History of Panama*, D'Arcy, W. G., and M. D. Correa (eds.). St. Louis: Missouri Botanical Garden.

Croat, T. B. 1978. *Flora of Barro Colorado Island*. Stanford, CA: Stanford University Press.

Darwin, C. 1867. On the movements and habits of climbing plants. *J. Linn. Soc., Bot.* 9: 1–118.

Emmons, L. H., and A. H. Gentry. 1983. Tropical forest structure and the distribution of gliding and prehensile-tailed vertebrates. *Amer. Naturalist* 121: 513–524.

Gentry, A. H. 1986. Endemism in tropical versus temperate plant communities. Pp. 153–181 in *Conservation Biology: The Science of Scarcity and Diversity*, Soulé, M. D. (ed.). Sunderland, MA: Sinauer Associates.

___. 1988. Changes in plant community diversity and floristic composition on environmental and geographical gradients. *Ann. Missouri Bot. Gard.* 75: 1–34.

___. 1991. The distribution and evolution of climbing plants. Pp. 3–42 in *The Biology of Vines*, Putz, F. E., and H. A. Mooney (eds.). New York: Cambridge University Press.

___. 1995. Patterns of diversity and floristic composition in neotropical montane forests. Pp. 103–126 in *Biodiversity and Conservation of Neotropical Montane Forests*, Churchill, S. P., H. Balslev, E. Forero, and J. L. Luteyn (eds.). New York: New York Botanical Garden.

Harris, L. D., and G. Silva-Lopez. 1992. Forest fragmentation and the conservation of biological diversity. Pp. 197–237 in *Conservation Biology: The Theory and Practice of Nature Conservation, Preservation, and Management*, Fiedler, P. L. and S. K. Jain (eds.). New York: Routledge, Chapman & Hall, Inc.

Hartshorn, G. S. 1983. Plants. Pp. 118–157 in *Costa Rican Natural History*, Janzen, D. (ed.). Chicago: University of Chicago Press.

Holdridge, L. R., W. C. Grenke, W. H. Hatheway, T. Liang, and J. A. Tosi, Jr. 1971. *Forest Environments in Tropical Lifezones: A Pilot Study*. New York: Pergamon Press.

Jacobs, M. 1976. The study of lianas. *Fl. Males. Bull.* 29: 2610–2618.

Klein, R. M. 1987. *The Green World: An Introduction to Plants and People*. New York: Harper Collins Publishers.

Krukoff, B. A. 1965. Supplementary notes on the American species of *Strychnos*, VII. *Mem. New York Bot. Gard.* 12: 1–94.

Montgomery, G., and M. Sunquist. 1978. Habitat selection and use by two-toed and three-toed sloths. Pp. 329–359 in *Ecology of Arboreal Folivores*, Montgomery, G. (ed.). Washington, DC: Smithsonian Institute Press.

Peñalosa, J. 1984. Basal branching and vegetative spread in two tropical rain forest lianas. *Biotropica* 16: 1–9.

Phillips, O. 1991. The ethnobotany and economic botany of tropical vines. Pp. 427–475 in *The Biology of Vines*, Putz, F. E., and H. A. Mooney (eds.). New York: Cambridge University Press.

Plotkin, M. J. 1993. *Tales of a Shaman's Apprentice*. New York: Penguin Group.

Putz, F. E. 1982. *Natural History of Lianas and Their Influences on Tropical Forest Dynamics*. Unpublished Ph.D. thesis. Ithaca, NY: Cornell University.

___. 1984. The natural history of lianas on Barro Colorado Island, Panama. *Ecology* 65: 1713–1724.

Raven, P. H. 1976. Ethics and attitudes. Pp. 155–179 in *Conservation of Threatened Plants*, Simmons, J., et al. (eds.). New York: Plenum Press.

Reis, S. von, and F. J. Lipp, Jr. 1982. *New Plant Sources for Drugs and Foods from the New York Botanical Garden*. Boston: Harvard University Press.

Schultes, R. E., and A. Hofmann. 1979. *Plants of the Gods: Origins of Hallucinogenic Use*. New York: McGraw-Hill Publications.

Walter, H. 1985. *Vegetation of the Earth and Ecological Systems of the Geo-biosphere*. New York: Springer Verlag.

Wilson, E. O. 1988. The current state of biological diversity. Pp. 3–18 in *Biodiversity*, Wilson, E. O. (ed.). Washington, DC: National Academy Press.

Chapter 3

Croat, T. B. 1978. *Flora of Barro Colorado Island*. Stanford, CA: Stanford University Press.

Gentry, A. H. 1973. Flora of Panama: Family 172, Bignoniaceae. *Ann. Missouri Bot. Gard.* 60: 781–977.

___. 1974a. Flowering phenology and diversity in tropical Bignoniaceae. *Biotropica* 6: 64–68.

___. 1974b. Coevolutionary patterns in Central American Bignoniaceae. *Ann. Missouri Bot. Gard.* 61: 728–759.

___. 1975. *Bignonia crucigera*: A case of mistaken identity. *Taxon* 24: 121–123.

___. 1977. New species of Leguminosae, Lauraceae, and Monimiaceae and new combinations in Bignoniaceae from western Ecuador. *Selbyana* 2: 39–45.

___. 1982. Pattern of neotropical plant species diversity. *Evol. Bio.* 15: 1–84.

___. 1985. An ecotaxonomic survey of Panamanian lianas. Pp. 29–42 in *The Botany and Natural History of Panama*, D'Arcy, W. G., and M. D. Correa (eds.). Monogr. Syst. Bot. Missouri Bot. Gard.

———. 1991. Distribution and evolution of climbing plants. Pp. 3–42 in *The Biology of Vines*, Putz, F. E., and H. A. Mooney (eds.). New York: Cambridge University Press.

———. 1993. *A Field Guide to the Families and Genera of Woody Plants of Northwest South America*. Washington, DC: Conservation International.

Richardson, I. B. K. 1978. Bignoniaceae. Pp. 249–250 in *Flowering Plants of the World*, Heywood, V. H., D. M. Moore, I. B. K. Richardson, and W. T. Stearn (eds.). New York: Mayflower Books.

Standley, P. C. 1930. Flora of Yucatan. *Publ. Field Mus. Nat. Hist., Bot. Ser.* 3: 417–421.

Standley, P. C., and S. Calderón. 1925. *Lista preliminar de las plantas de El Salvador*. San Salvador, El Salvador: Tip. la Unión, Dutriz hermanos.

Chapter 4

Condon, M. A., and L. E. Gilbert. 1988. Sex expression of *Gurania* and *Psiguria* (Cucurbitaceae): Neotropical vines that change sex. *Amer. J. Bot.* 75: 875–884.

Dieterle, J. V. A. 1976. Flora of Guatemala: Cucurbitaceae. *Fieldiana, Bot.* 24: 306–389.

Haber, W. A. 1991. Lista provisional de las plantas de Monteverde. *Brenesia* 34: 63–120.

Hammel, B. E., and W. G. D'Arcy. 1981. New taxa from the uplands of western Panama. *Ann. Missouri Bot. Gard.* 68: 213–217.

Hampshire, R. J. 1992a. *Gurania sessiliflora* (Cucurbitaceae), a new species from Panama. *Novon* 2: 201–203.

———. 1992b. A new combination in *Psiguria* (Cucurbitaceae) from Mesoamerica. *Novon* 2: 339–340.

Inbio (Institute Nacional de Biodiversidad). 1998. *Manual de Plantas de Costa Rica*. Santo Domingo de Heredia, Costa Rica: Inst. Nac. de Biodiversidad.

Jeffrey, C. 1971. Further notes on Cucurbitaceae: II, the tribe Cucurbiteae. *Kew Bull.* 25: 191–236.

———. 1978. Further notes on Cucurbitaceae: IV, some new world taxa. *Kew Bull.* 33: 347–380.

———. 1990. An outline of Cucurbitaceae. Pp. 449–463 in *Biology and Utilization of the Cucurbitaceae*, Bates, D. M., R. W. Robinson, and C. Jeffrey (eds.). Ithaca, NY: Cornell University Press.

Kearns, D. M. 1992. A revision of *Sechiopsis* (Cucurbitaceae). *Sys. Bot.* 17: 395–408.

Lira, R. 1995. A new species of *Sicydium* Schlechtendal (Cucurbitaceae: Zanonioideae, Zanonieae, Sicydiinae) for the Flora Mesoamericana. *Novon* 5: 284–286.

Lira, R., and F. Chiang. 1992. Two new combinations in *Sechium* (Cucurbitaceae) from Central America, and a new species from Oaxaca, Mexico. *Novon* 2: 227–231.

Nee, M. 1993. Cucurbitaceae. Flora de Veracruz, fasc. 74. Instituto Ecologia, A.C., Xalapa, Veracruz, Mexico.

Wilbur, R. L., and collaborators. 1994. Vascular plants: An interim checklist. Pp. 350–378 in *La Selva: Ecology and Natural History of a Neotropical Rainforest*, McDade, L. A., K. S. Bawa, H. A. Hespensheide, and G. S. Hartshorn (eds.). Chicago, IL: University of Chicago Press.

Wunderlin, R. P. 1976. Two new species and a new combination in *Frantzia* (Cucurbitaceae). *Brittonia* 28: 239–244.

———. 1978. Flora of Panama: Cucurbitaceae. *Ann. Missouri Bot. Gard.* 65: 285–368.

Chapter 5

Chant, S. R. 1978. Leguminosae. Pp. 149–152 in *Flowering Plants of the World*, Heywood, V. H., D. M. Moore, I. B. K. Richardson, and W. T. Stearn (eds.). New York: Mayflower Books.

Croat, T. B. 1978. *Flora of Barro Colorado Island*. Stanford, CA: Stanford University Press.

Gentry, A. H. 1991. Distribution and evolution of climbing plants. Pp. 3–42 in *The Biology of Vines*, Putz, F. E., and H. A. Mooney (eds.). New York: Cambridge University Press.

Standley, P. C., and J. Steyermark. 1946. Flora of Guatemala: *Vicia*. *Fieldiana, Bot.* 24: 361–363.

Woodson, R. E., Jr., and R. W. Schery. 1950. Flora of Panama: *Entada*. *Ann. Missouri Bot. Gard.* 37: 309–312.

Wunderlin, R. P. 1976. The Panamanian species of *Bauhinia* (Leguminosae). *Ann. Missouri Bot. Gard.* 63: 346–354.

Chapter 6

Blackwell, W. H., Jr. 1967. Family 159. Loganiaceae. *Ann. Missouri Bot. Gard.* 54: 393–413.

Ducke, A. 1955. O genero *Strychnos* no Brasil. *Boletim Tecnico Instituto Agronomico do Norte* 30: 1–64.

Jones, S. W. 1978. Loganiaceae. Pp. 222–223 in *Flowering Plants of the World*, Heywood, V. H., D. M. Moore, I. B. K. Richardson, and W. T. Stearn (eds.). New York: Mayflower Books.

Krukoff, B. A. 1965. Supplementary notes on the American species of *Strychnos*, VII. *Mem. New York Bot. Gard.* 12: 1–94.

Krukoff, B. A., and J. Monachino. 1942. The American species of *Strychnos*. *Brittonia* 4: 248–322.

Krukoff, B. A., and R. C. Barneby. 1969. Supplementary notes on the American species of *Strychnos*, VIII. *Mem. New York Bot. Gard.* 20: 1–93.

___. 1971. Supplementary notes on the American species of *Strychnos*, XI. *Phytologia* 22: 225–243.

Leeuwenberg, A. J. M. 1969. The Loganiaceae of Africa VIII, *Strychnos* III. *Mededelingen Landbouwhogeschool Wageningen* 69: 1–316.

Reis, S. von, and F. J. Lipp, Jr. 1982. *New Plant Sources for Drugs and Foods from the New York Botanical Garden*. Boston, MA: Harvard University Press.

Chapter 7

Holm-Nielsen, L. B., P. M. Jørgensen, and J. E. Lawesson. 1988. *Passifloraceae. Flora of Ecuador, no. 31*. Department of Botany, University of Goeteborg, and Section for Botany, Riksmuseum, Stockholm, Sweden.

Killip, E. P. 1938. The American species of Passifloraceae. *Publ. Field Mus. Nat. Hist., Bot. Ser.* 19: 1–613.

Chapter 8

Austin, D. F. 1971. Flora of Panama: Polemoniaceae. *Ann. Missouri Bot. Gard.* 58: 355–361.

Gibson, D. N. 1970. Flora of Guatemala: *Cobaea*. *Fieldiana, Botany* 24: 88–93.

Moore, D. M. 1978. Polemoniaceae. Pp. 233 in *Flowering Plants of the World*, Heywood, V. H., D. M. Moore, I. B. K. Richardson, and W. T. Stearn (eds.). New York: Mayflower Books.

Prather, L. A. 1999. Systematics of *Cobaea* (Polemoniaceae). *Syst. Bot. Monogr.* 57: 1–81.

Standley, P. C. 1914. A revision of the genus *Cobaea*. *Contr. U.S. Natl. Herb.* 17: 448–458.

Chapter 9

Burger, W. 1983. Family 62: Polygonaceae. *Fieldiana, Botany* 13: 99–138.
Wright, T. J. 1979. Polygonaceae. Pp. 77–78 in *Flowering Plants of the World*, Heywood, V. H., D. M. Moore, I. B. K. Richardson, and W. T. Stearn (eds.). New York: Mayflower Books.

Chapter 10

Howard, R. A. 1979. *Flora of the Lesser Antilles*. Jamaica Plain, MA: Arnold Arboretum, Harvard University.
Jones, S. W. 1978. Rhamnaceae. Pp. 187–188 in *Flowering Plants of the World*, Heywood, V. H., D. M. Moore, I. B. K. Richardson, and W. T. Stearn (eds.). New York: Mayflower Books.
Standley, P. C., and J. A. Steyermark. 1949. Flora of Guatemala: Rhamnaceae. *Fieldiana, Bot. Ser.* 24: 277–293.

Chapter 11

Croat, T. B. 1976. Flora of Panama: Sapindaceae. *Ann. Missouri Bot. Gard.* 63: 419–540.
___. 1978. *Flora of Barro Colorado Island*. Stanford, CA: Stanford University Press.
Morales, J. F. 2000. Manual de al Flora de Costa Rica: Sapindaceae (first draft). http://www.mobot.org/MOBOT/research/treat/sapind.shtml.
Morley, B. 1978. Sapindaceae. Pp. 193–194 in *Flowering Plants of the World*, Heywood, V. H., D. M. Moore, I. B. K. Richardson, W. T. Stearn (eds.). New York: Mayflower Books.
Radlkofer, L. 1931–1934. Sapindaceae. In *Das Pflanzenreich*, A. Engler, 98: 1–1539.
Standley, P. C., and J. A. Steyermark. 1949. Sapindaceae. *Fieldiana, Bot. Ser.* 24: 234–273.

Chapter 12

Chant, S. R. 1978. Smilacaceae. P. 320 in *Flowering Plants of the World*, Heywood, V. H., D. M. Moore, I. B. K. Richardson, and W. T. Stearn (eds.). New York: Mayflower Books.
Croat, T. B. 1978. *Flora of Barro Colorado Island*. Stanford, CA: Stanford University Press.
Haber, W. A. 1991. Lista provisional de las plantas de Monteverde, Costa Rica. *Brenesia* 34: 63–120.
Huft, M. J. 1994. Smilacaceae. Pp. 20–25 in *Flora Mesoamericana*, Vol. 6. G. Davidse, M. Sousa S., and A. O. Cheter (eds.). Univ. Nac. Auton Mex./Missouri Botanical Garden/Nat. Hist. Mus. (London).
Killip, E. P., and C. V. Morton. 1936. A revision of the Mexican and Central American species of *Smilax*. *Publ. Carnegie Inst. Wash.* 461: 255–296.
Morton, C. V. 1945. Smilacaceae. *Ann. Missouri Bot. Gard.* 32: 6–11.
___. 1962. A reexamination of Mexican *Smilax* (Smilacaceae). *Brittonia* 14: 299–309.
Standley, P. C. 1937. Smilacaceae. *Publ. Field Mus. Nat. Hist., Bot. Ser.* 18: 171–174.
Woodson, R. E., and R. W. Schery. 1942. Contributions toward a flora of Panama. VI. *Ann. Missouri Bot. Gard.* 29:326.

Chapter 13

Acevedo-Rodriguez, P., and R. O. Woodbury. 1985. *Los Bejucos de Puerto Rico: Volumen 1.* USDA: Forest Service, General Technical Report SO-58.

Chant, S. R. 1978. Vitaceae. Pp. 188–189 in *Flowering Plants of the World*, Heywood, V. H., D. M. Moore, I. B. K. Richardson, and W. T. Stearn (eds.). New York: Mayflower Books.

Croat, T. B. 1973. A new species of *Cissus* (Vitaceae) for Central and South America. *Ann. Missouri Bot. Gard.* 60: 564–567.

___. 1978. *Flora of Barro Colorado Island*. Stanford, CA: Stanford University Press.

Dodson, C. H., A. H. Gentry, and F. M. Valverde. 1985. *La Flora de Jauneche, Ecuador*. Quito: Banco Central de Ecuador.

Dodson, C. H., and A. H. Gentry. 1978. Flora of the Rio Palenque Science Center. *Selbyana* 4: 1–628.

Elias, T. S. 1968. Flora of Panama: Vitaceae. *Ann. Missouri Bot. Gard.* 55: 81–92.

Standley, P. C. 1930. Flora of Yucatan. *Publ. Field Mus. Nat. Hist., Bot. Ser.* 3: 342–343.

Wunderlin, R. P. 1982. *Guide to the Vascular Plants of Central Florida*. Tampa: University Presses of Florida.

INDEX

C

P

Printed and bound by CPI Group (UK) Ltd, Croydon, CR0 4YY
16/09/2026
14963448-0003